A Brief Atlas of the

HUMAN BODY

Matt Hutchinson
Jon Mallatt
Elaine N. Marieb
Patricia Brady Wilhelm

Photographs by
Ralph T. Hutchings and Nina Zanetti

PEARSON

Benjamin
Cummings

San Francisco Boston New York
Cape Town Hong Kong London Madrid Mexico City
Montreal Munich Paris Singapore Sydney Tokyo Toronto

Editor-in-Chief: Serina Beauparlant
Project Editor: Karoliina Tuovinen
Assistant Editor: Jessica Brunner
Managing Editor, Production: Wendy Earl
Production Manager: Janet Vail
Cover and Front Matter Designer: Yvo Riezebos Design

Photography Credits: Figures 1 – 71*: Ralph T. Hutchings
Plates 1 – 55: Nina Zanetti, Benjamin Cummings
Publishers, Pearson Education

*Except Figure 57 from *The Bassett Atlas of Human Anatomy* by Robert A. Chase,
©1989, The Benjamin/Cummings Publishing Company, Inc.,
and Figure 68a photographed by John Martinek, Kirkwood Community College.

ISBN 0-8053-7373-X
26 27 28 29 30 31 32 33 34 35 —DOM— 17 16 15

Pearson Benjamin Cummings is a trademark of
Pearson Education, Inc.
www.aw-bc.com

PREFACE

Building upon our original vision, the second edition of *A Brief Atlas of the Human Body* features 51 soft tissue images and 104 bone images providing a degree of clarity and scale that could never be achieved within a textbook alone. In addition, this edition includes a brand new section, histology of basic tissues and select organs, containing 55 outstanding histology slides photographed by Nina Zanetti of Siena College.

Elaine Marieb chose many of the soft tissue views. Matt Hutchinson, of Washington State University, took on the arduous task of labeling each structure. Jon Mallatt, co-author of the human anatomy text, scrutinized and approved the views, leaders, and labels. References to related atlas images herein can be found in the illustration figure legends of both the human anatomy and the human anatomy and physiology textbooks.

For this edition, Patricia Brady Wilhelm, co-author of the human anatomy text, reviewed each photograph for accuracy and chose several new soft tissue images from the collection of Mark Nielsen and Shawn Miller of the University of Utah. She also worked to choose all of the slides included in the histology portion of the Atlas to represent the most useful selection of histology plates for students in the classroom.

Ralph T. Hutchings, formerly with The Royal College of Surgeons of England, photographed each of the bone structures and many of the soft tissue images found in this book. His reputation as an anatomical photographer preceded him, and we certainly were not disappointed—the quality of his work is here for all to see. We are most grateful to him for lending his expertise to this project, and for his good humor and ready willingness to meet our demands. We are grateful to John Martinek of Kirkwood Community College, who contributed his excellent photograph of the internal surface of the stomach (Figure 69a).

The authors would like to thank the following instructors for their expertise and thoughtful feedback in reviewing the Atlas: Andy Beall, University of North Florida; Dennis Carnes, Imperial Valley College; Leslie Hendon, The University of Alabama at Birmingham; H. Rodney Holmes, Waubonsee Community College; Jeff Kent, Volunteer State Community College; Mark Robertson, Delta College; Laura Rosillo, IVY Tech State College; Justine Wilcox, University of North Florida; Joseph Yavornitzky, Baldwin Wallace College; Hillman Mann, Volunteer State Community College; Darrell Davies, Kalamazoo Valley Community College; Louise Russo, Villanova University; Tom Swensen, Ithaca College.

We are hopeful that the second edition of *A Brief Atlas of the Human Body* proves to be a relevant source to students and instructors. Benjamin Cummings would welcome your comments and suggestions, which may be sent to the following address:

Publisher
Applied Sciences
Benjamin Cummings
1301 Sansome Street
San Francisco, CA 94111

CONTENTS

Part II   BONES OF THE HUMAN SKELETON   25

Part III   SOFT TISSUE OF THE HUMAN BODY   83

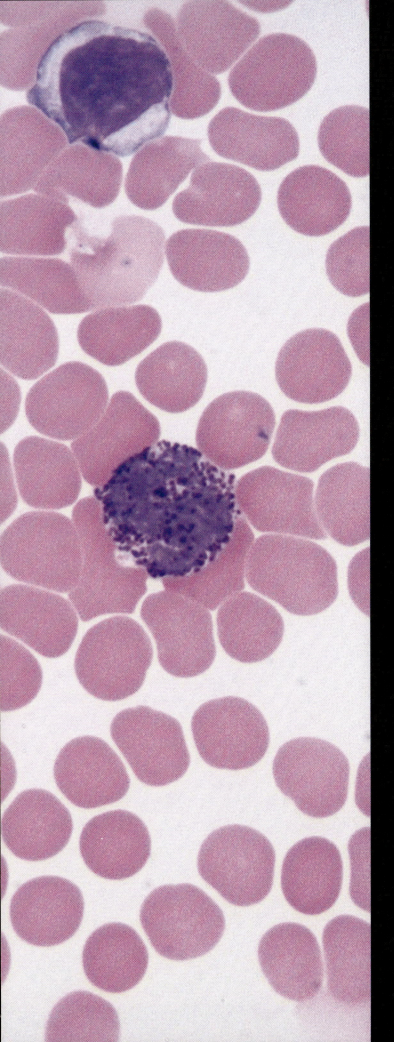

Part I

HISTOLOGY

BASIC TISSUES

Cell nucleus

Cytoplasm

Plasma membrane

PLATE 1

Simple squamous epithelium, surface view. Silver stained mesothelium (360×)

DESCRIPTION:

Single layer of flat cells with disc shaped central nuclei and little cytoplasm. In this surface view, the cells resemble fried eggs.

LOCATION:

Form the kidney glomeruli and corpuscles; air sacs (alveoli) of the lungs; lining of the heart, blood vessels and lymphatic vessels; lining of the ventral body cavity (serosae).

Simple squamous epithelium

Cell nucleus

PLATE 2

Simple squamous epithelium, section through a renal corpuscle in the renal cortex (465×)

DESCRIPTION:

The single layer of flat cells with disc shaped central nuclei and small amount of cytoplasm are apparent in the parietal layer of the renal capsule. In this section, the thinness of the squamous cells and the dark staining central nuclei are obvious.

LOCATION:

As listed for Plate 1.

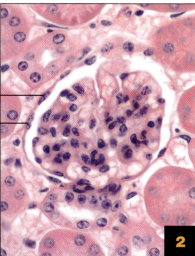

Simple cuboidal epithelium

Lumen of renal tubule

Epithelial cell nucleus

PLATE 3

Simple cuboidal epithelium, l.s. through renal medulla (350×)

DESCRIPTION:

Single layer of cube-shaped cells with large round central nuclei. This section shows multiple rows of simple cuboidal epithelium forming the renal tubules.

LOCATION:

Forms the kidney tubules and collecting ducts; the ducts and secretory portions of many glands; and the surface of the ovary.

Simple columnar epithelium

Nucleus

Microvilli

Goblet cell secreting mucus

Lamina propria

Ciliated simple columnar epithelium

Nucleus

Lumen of uterine tube

Cilia

Pseudostratified columnar epithelium

Nuclei

Goblet cell

Cilia

PLATE 4 — Non-ciliated simple columnar epithelium from the small intestine—jejunum (360×)

DESCRIPTION: Single layer of column shaped cells with either a round or oval shaped nucleus. Unicellular glands (goblet cells) that secrete mucous are common in this tissue. Microvilli, extensions of the plasma membrane of the apical surface, are present in the small intestine.

LOCATION: Lines digestive tract from stomach to anal canal; the gallbladder; portions of uterus and uterine tubes; and the excretory ducts of some glands.

PLATE 5 — Ciliated simple columnar epithelium from the oviduct (350×)

DESCRIPTION: Single layer of column shaped cells with either an oval or a round nucleus with cilia extending from the apical surface.

LOCATION: Lines the small bronchi, the uterine tubes, and portions of the uterus.

PLATE 6 — Pseudostratified columnar epithelium from the trachea (335×)

DESCRIPTION: Single layer of cells of differing heights, some not reaching the apical surface. Nuclei located at different levels give the appearance of a multilayered (stratified) tissue. Mucous secreting goblet cells are common in this tissue. There are both ciliated and non-ciliated types. Goblet cells and cilia are seen in this specimen.

LOCATION: The non-ciliated type lines the sperm carrying ducts (duct of epididymis, ductus deferens, ejaculatory duct, and mid-portion of the male urethra); the ciliated type lines the trachea and most of the upper respiratory tract.

Apical cells

Stratified squamous epithelium

Basal cells

Cell nuclei

Connective tissue

7

PLATE 7

Stratified squamous epithelium from the mucosa of the esophagus (350×)

DESCRIPTION:

Distinguished by multiple layers of cells with nuclei distributed throughout. The basal cells are cuboidal or columnar in shape, are metabolically active and have a high rate of mitosis; the apical cells are flat (squamous). In the keratinized type the surface cells are filled with the protein keratin.

LOCATION:

The non-keratinized type (seen in this specimen) lines the esophagus, mouth, vagina and anus. The keratinized type forms the epidermis of the skin.

Stratified cuboidal epithelium

Lumen of duct

Cell nuclei

Connective tissue

Sweat gland

8

PLATE 8

Stratified cuboidal epithelium from the duct of a sweat gland (340×)

DESCRIPTION:

Two layers of cells, identifiable by the two rows of round nuclei. The apical layer is composed of cube shaped cells.

LOCATION:

Lines the largest ducts of sweat glands, the mammary glands, and the salivary glands.

Stratified columnar epithelium

Columnar apical cell

Cuboidal basal cell

Cell nuclei

Lumen of duct

Connective tissue

9

PLATE 9

Stratified columnar epithelium from a duct in the parotid gland (350×)

DESCRIPTION:

Two or more layers of cells: the apical layer is composed of columnar shaped cells, the basal layer is usually cuboidal cells.

LOCATION:

Lines the large ducts of some glands, and some portions of the male urethra.

5

Lumen of ureter

Cuboidal shaped cells

Transitional epithelium

Connective tissue

10

Nuclei of fibroblasts

Collagen fiber

Elastic fibers

Gel-like ground substance

Mast cell

11

Cell nucleus of an adipocyte

Adipocytes

Fat droplet within an adipocyte

Blood vessel

12

EPITHELIAL TISSUES, *continued*

PLATE 10   Transitional epithelium from the ureter (340×)

DESCRIPTION: As its name implies, this tissue changes shape. It is composed of multiple layers of cells. When in relaxed state, as shown here, the cells appear cuboidal in shape; when stretched, the cells appear squamous in shape.

LOCATION: Forms the lining of the urinary bladder, ureter, and the superior portion of the urethra.

CONNECTIVE TISSUES— CONNECTIVE TISSUE PROPER

PLATE 11   Areolar connective tissue (350×)

DESCRIPTION: Matrix contains all three fiber types (collagen fibers, elastic fibers, and reticular fibers) within a gel-like ground substance. Fibroblasts, mast cells, macrophages and other white blood cells are found within this tissue.

LOCATION: Distributed throughout the body loosely binding adjacent structures: forms the lamina propria that underlies all epithelia in the body; forms the papillary layer of the dermis of the skin and contributes to the superficial fascia; surrounds blood vessels, nerves, muscles.

PLATE 12   Adipose tissue from the external ear (350×)

DESCRIPTION: Distinguished by the closely packed adipocytes (fat cells) within a sparse matrix. Each adipocyte is filled with a large fat droplet causing the nucleus to be pushed to the edge of the cell.

LOCATION: A ubiquitous tissue found throughout the body: the hypodermis of the skin; surrounding the kidneys, eyeballs, mammary glands, and many other body organs; within the abdomen; and within the fascial planes separating muscle layers.

Reticular fibers

Gel-like ground substance

PLATE 13

Reticular connective tissue, lymph node (350×)

DESCRIPTION:

Matrix is composed of reticular fibers loosely distributed within a gel-like ground substance. Cellular components are fibroblasts, lymphocytes, and other blood cells.

LOCATION:

Forms the internal framework of many lymphoid organs: spleen, lymph nodes, bone marrow.

Nuclei of fibroblasts

Collagen fibers

PLATE 14

Dense irregular connective tissue from the submucosa of the large intestine (350×)

DESCRIPTION:

Distinguished by the irregular arrangement of fibers densely packed in multiple directions. Composed primarily of collagen fibers with some elastic fibers. Major cell type is the fibroblast.

LOCATION:

Reticular layer of dermis of skin, submucosa of digestive tract, fibrous capsules of organs and joints.

Regularly aligned collagen fibers

Nuclei of fibroblasts

PLATE 15

Dense regular connective tissue, tendon (340×)

DESCRIPTION:

Densely packed fibers, primarily collagen, arranged parallel to each other. The nuclei of the fibroblasts are also aligned in parallel. This is an important feature for differentiating this tissue from smooth muscle tissue.

LOCATION:

Tendons, most ligaments, aponeuroses.

7

Lumen of aorta

Elastic fibers

16

Matrix

Chondrocyte in a lacuna

Lacuna

Perichondrium

17

Chondrocyte in a lacuna

Elastic fibers

Gelatinous ground substance

18

CONNECTIVE TISSUES—PROPER, *continued*

PLATE 16   Elastic connective tissue from the aorta (90×)

DESCRIPTION: Connective tissue with densely packed elastic fibers. Notice the wavy appearance of the dark staining elastic fibers. Here, these fibers are within the smooth muscle layer of the wall of the aorta.

LOCATION: Found within the body wall of arteries; distributed throughout the trachea and bronchial tree; located within some ligaments in the neck region.

CONNECTIVE TISSUES—CARTILAGE

PLATE 17   Hyaline cartilage from the trachea (320×)

DESCRIPTION: Cartilage cells (chondrocytes) located within spaces (lacunae) in the tissue matrix. The matrix is a firm, gel-like ground substance embedded with collagen fibrils, which are not viewable via light microscopy. Vascularized perichondrium surrounds the cartilage nourishing the tissue and producing new tissue.

LOCATION: Forms most of the embryonic skeleton; covers the ends of long bones in joint cavities; forms the costal cartilages, the cartilages of the nose, trachea and bronchial tree, and most of the laryngeal cartilages.

PLATE 18   Elastic cartilage (350×)

DESCRIPTION: As in hyaline cartilage, the chondrocytes sit in spaces (lacunae) within the tissue matrix. The matrix contains a firm gel-like ground substance and both collagen fibrils and elastic fibers. In this preparation the elastic fibers are the dark purple strands viewable in the matrix.

LOCATION: Epiglottis and the external ear.

Chondrocyte in a lacuna

Collagen fibers

PLATE 19   Fibrocartilage, within a tendon (350×)

DESCRIPTION: The gelatinous matrix is densely packed with thick collagen fibers. Chondrocytes are located in lacunae, spaces in the matrix. This feature distinguishes this tissue from dense irregular connective tissue. This is the strongest type of cartilage.

LOCATION: Form the intervertebral discs; the pubic symphysis; the articular discs within joint cavities. Also located within some tendons, particularly where a tendon passes around a bony pulley.

CONNECTIVE TISSUES—BONE

Lamellae

Osteon

Central canal

Osteocytes in lacunae

Canaliculi extending through matrix

PLATE 20   Compact bone (90×)

DESCRIPTION: Tissue composed of a hard, calcified matrix containing many collagen fibers. This densely packed bone tissue is organized in lamellae (layers of bone tissue) and osteons (concentric rings of bone tissue). Blood vessels are located in the central canals; osteocytes lie in the lacunae; canaliculi, the thin dark lines, connect adjacent osteocytes.

LOCATION: Found in the shaft of long bones; the external portion of flat, short, and irregular shaped bones; and the external portion of the epiphyses.

PLATE 21   Spongy bone (340×)

DESCRIPTION: Composed of the same materials (hard calcified ground substance and collagen fibers) as compact bone. Arranged into small beams (trabeculae) of bony tissue. Spaces between trabeculae are filled with red bone marrow. Bone arranged in lamellae and osteocytes are located in lacunae. Bone forming cells, osteoblasts, line the trabeculae. Bone destroying cells, osteoclasts, also present as spongy bone continually remodels.

LOCATION: Located in the internal regions of the epiphyses, as well in the internal portions of flat, short, and irregular shaped bone.

Red bone marrow

Trabecula

Osteocytes within lacunae

Osteoblasts

Osteoclasts

Erythrocytes

Platelets

Leukocytes
(neutrophils)

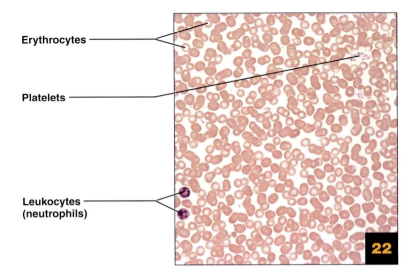

PLATE 22   Blood smear (270×)

DESCRIPTION: Erythrocytes (red blood cells), leukocytes (white blood cells), and platelets in a fluid matrix. Red blood cells, the small, red, disc-shaped cells shown here, are the most numerous formed element in blood.

LOCATION: Contained within blood vessels.

Neutrophils

Multilobed nucleus

PLATE 23   Blood smear, neutrophils (895×)

DESCRIPTION: Granular leukocyte containing a multilobed nucleus. Cytoplasm appears light purple due to stain affinities. This is the most common leukocyte.

LOCATION: Originate and are stored in the red bone marrow. Travel in blood vessels. Enter loose connective tissues in response to infections. Respond to bacterial infections.

Eosinophil

Bilobed nucleus

PLATE 24   Blood smear, eosinophil (885×)

DESCRIPTION: Granular leukocyte with a bilobed nucleus, cytoplasmic granules stain red. Relatively rare (1–4% of all leukocytes).

LOCATION: Originate and are stored in the red bone marrow. Travel in blood vessels. Enter loose connective tissues in response to infections. Respond to parasitic infection, and to end an allergic reaction.

Basophil
Cytoplasmic granules

PLATE 25 Blood smear, basophil (840×)

DESCRIPTION: Granular leukocyte with a bilobed nucleus, cytoplasmic granules stain dark purple. Most rare (0–1% of all leukocytes).

LOCATION: Originate and are stored in the red bone marrow. Travel in blood vessels. Enter loose connective tissues in response to infections. Mediate inflammation by secreting histamines.

Lymphocyte

Nucleus

Cytoplasm

PLATE 26 Blood smear, lymphocytes (815×)

DESCRIPTION: Agranular leukocyte with a large circular nucleus that takes up most of the cell volume and stains purple, surrounded by a thin border of pale blue cytoplasm.

LOCATION: Originate in red bone marrow. Travel in blood vessels. Enter loose connective tissues and lymphoid tissue in response to infections.

Monocyte

Nucleus

PLATE 27 Blood smear, monocyte (855×)

DESCRIPTION: Largest leukocyte; agranular with a kidney shaped nucleus that stains lighter than that of lymphocytes. Contains a larger amount of blue staining cytoplasm than in lymphocytes.

LOCATION: Originate and are stored in the red bone marrow. Travel in blood vessels. Enter loose connective tissues in response to infections, transform into macrophages to phagocytize foreign matter.

11

Muscle fiber

Nuclei

A band
I band
Z disc

28

PLATE 28   Skeletal muscle l.s. (445×)

DESCRIPTION: Long cylindrical cells (fibers), multinucleated, obvious striations running perpendicular to fiber direction. Dark bands are called A bands, light bands are I bands. In this section you can also see the Z discs, the thin dark lines running through the middle of the I bands.

LOCATION: In skeletal muscles.

Perimysium
surrounding
a fascicle

Blood vessels

Muscle fibers

Nuclei of
muscle fiber

29

PLATE 29   Skeletal muscle c.s. (85×)

DESCRIPTION: Cross section through skeletal muscle showing muscle fibers and connective tissues. Nuclei are located at the periphery of each fiber and the connective tissue, perimysium, (stained gray) groups bundles of muscle fibers into fascicles. Notice the blood vessels running within the perimysium.

LOCATION: In skeletal muscles.

Skeletal muscle fiber
Axon of motor neuron

Axon terminals at
neuromuscular
junctions

Terminal branches

30

PLATE 30   Neuromuscular junction (motor end plate; 290×)

DESCRIPTION: Junction of a motor neuron with skeletal muscle fibers. The axon branches into multiple axon terminals that innervate muscle fibers.

LOCATION: Within skeletal muscles.

MUSCLE TISSUE—CARDIAC MUSCLE

Nucleus

Striations

Intercalated discs

Cardiac muscle cell

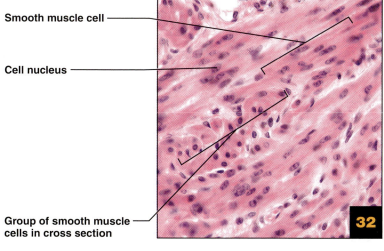

PLATE 31   Cardiac muscle (365×)

DESCRIPTION: Striated muscle (although striations are often difficult to view at this magnification) composed of branching cells with one centrally located nucleus (occasionally two). Cells are joined by specialized cell junctions, intercalated discs.

LOCATION: Makes up the myocardium of the heart.

MUSCLE TISSUE—SMOOTH MUSCLE

Smooth muscle cell

Cell nucleus

Group of smooth muscle cells in cross section

PLATE 32   Smooth muscle, from the uterus (300×)

DESCRIPTION: Non-striated muscle tissue. Elongated, tapering cells with a single, central nucleus are closely packed together to form sheets. Distinguishable from dense regular connective tissue because nuclei are randomly distributed throughout.

LOCATION: Composes the muscular layer in the wall of the digestive tract, circulatory vessels, urinary, and reproductive organs. Also located in the respiratory tubes and inside the eye.

NERVOUS TISSUE

Nucleus

Nucleolus

Cell processes

Neuroglia

PLATE 33   Neuronal cell body in central nervous system (350×)

DESCRIPTION: The neuronal cell body contains the nucleus and cellular organelles. Extending from the cell body are cell processes that either receive or transmit signals. The dark staining fibers are neurofibrils, intermediate filaments that run throughout the neuron. This image also shows multiple neuroglia, supportive cells that aid neuronal function.

LOCATION: Found in the central nervous system in the brain and spinal cord; in the peripheral nervous system in peripheral ganglia.

Myelinated axons

Endoneurium

Myelin sheath

Perineurium

Epineurium

34

Myelin sheath

Axon

Node of Ranvier

Nuclei of
Schwann cells

35

PLATE 34   Peripheral nerve, c.s. (260×)

DESCRIPTION: A nerve is composed of the axonal processes of numerous neurons. Myelin surrounds many of the axons. Connective tissues wrap the axons: the endoneurium surrounds each axon; the perineurium bundles groups of axons into fascicles; the epineurium covers the entire nerve.

LOCATION: Found throughout the body carrying sensory and motor innervation from/to the periphery.

PLATE 35   Peripheral nerve, l.s. (830×)

DESCRIPTION: Longitudinal section through a nerve showing multiple axons surrounded by myelin sheaths; Nodes of Ranvier, pinching of the myelin sheath that indicates the boundary between adjacent Schwann cells; and Schwann cell nuclei.

LOCATION: Found throughout the body innervating body organs.

SELECT ORGANS

Dura mater   Posterior median sulcus   Posterior funiculus

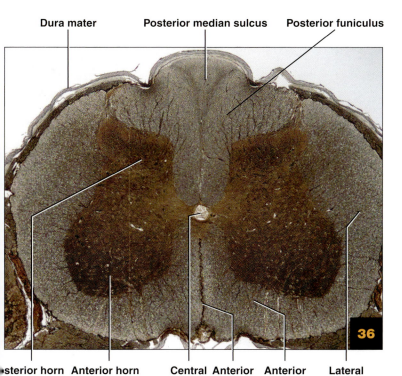

sterior horn Anterior horn   Central Anterior Anterior Lateral
                                    canal   median   funiculus funiculus
                                             fissure

PLATE 36

Spinal cord c.s. through lumbar region (18×)

DESCRIPTION:

The hollow central canal is surrounded by a butterfly-shaped region of gray matter forming the anterior and posterior horns (stained brown in this section) that contain neuronal cell bodies; short, non-myelinated interneurons; and neuroglia. The external white matter, the funiculi, (stained gray) contains myelinated axons that make up the ascending and descending spinal tracts.

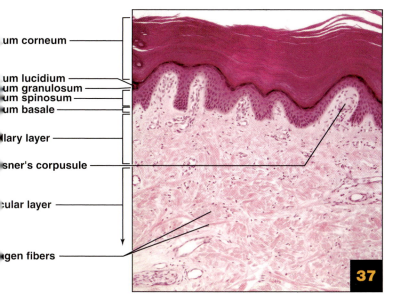

um corneum

um lucidium
um granulosum
um spinosum
um basale

llary layer

sner's corpusule

cular layer

gen fibers

PLATE 37

Thick skin showing epidermal and dermal regions (85×)

DESCRIPTION:

The epidermis, the dark pink region, is composed of the stratum basale, stratum spinosum, stratum granulosum, stratum lucidium, and the thick stratum corneum. Deep to the epidermis is the papillary layer of the dermis, composed of loose areolar connective tissue. Note the Meissner's corpuscle, a touch receptor, in the dermal papilla. The deepest layer of the dermis, the reticular layer, is composed of dense irregular connective tissue. Note the greater density of collagen fibers (stained pink) in this layer.

VEIN
- Lumen
- Tunica intima
- Tunica media
- Tunica externa

ARTERY
- Tunica intima
- Tunica media
- Elastic laminae
- Tunica externa

PLATE 38 Muscular artery and vein (80×)

DESCRIPTION: The vessel on the right, the artery, shows a wavy tunica intima resulting from the internal elastic lamina just deep to this layer. The thick tunica media is composed of multiple cell layers. Another layer of wavy elastic tissue, the external elastic lamina, is located outside of the tunica media. In the vein on the left, the lumen is irregular in shape, there are no elastic laminae, and the tunica media is only a few cells in thickness.

- Capsule
- Trabecula
- White pulp
- Arteriole
- Red pulp

PLATE 39 Spleen (17×)

DESCRIPTION: The outer capsule of the spleen is reticular connective tissue. Trabeculae, extensions of this tissue into the deeper portion of the spleen, provide a framework for the organ. Lymphoid tissue, containing B and T lymphocytes surrounding arterial branches forms the white pulp of the spleen. Surrounding these "islands" is the red pulp, composed of both venous sinuses and splenic cords.

- Mucosa
- Ciliated epithelium
- Lamina propria
- Submucosa
- Seromucosal gland
- Hyaline cartilage ring

PLATE 40 Trachea (355×)

DESCRIPTION: The mucosa, composed of pseudostratified ciliated columnar epithelium and the underlying lamina propria, lines the trachea. Note the columnar epithelial cells, multiple nuclei, and the distinctive cilia along the apical surface. External to this layer, the submucosa consists of connective tissue and imbedded seromucosal glands that secrete mucous. Cartilagenous rings, composed of hyaline cartilage make up the most external layer seen.

Respiratory bronchiole **Alveoli**

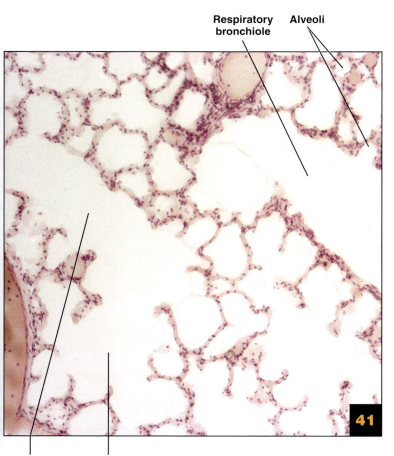

Alveolar duct **Alveolar sac**

PLATE 41 Lung (120×)

DESCRIPTION: This section through the lung shows the structures of the respiratory zone. The respiratory bronchiole, lined with simple cuboidal epithelium, has alveoli outpocketing from its wall. Alveolar ducts lead to alveolar sacs, terminal clusters of alveoli. Simple squamous epithelium forms the alveolar walls.

Simple columnar epithelium of stomach
Mucous cells
Gastric pit
Gastroesophageal transition
Stratified squamous epithelium of esophagus
Gastric glands

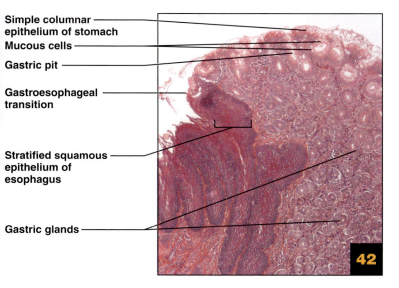

PLATE 42 Gastroesophageal junction (120×)

DESCRIPTION: The epithelial tissue changes abruptly at the gastroesophageal junction, from stratified squamous epithelium in the esophagus (on the left side of the image) to simple columnar epithelium of the stomach (top of the image). The gastric pits and gastric glands of the mucosal layer of the stomach, also composed of simple columnar epithelium, are apparent.

17

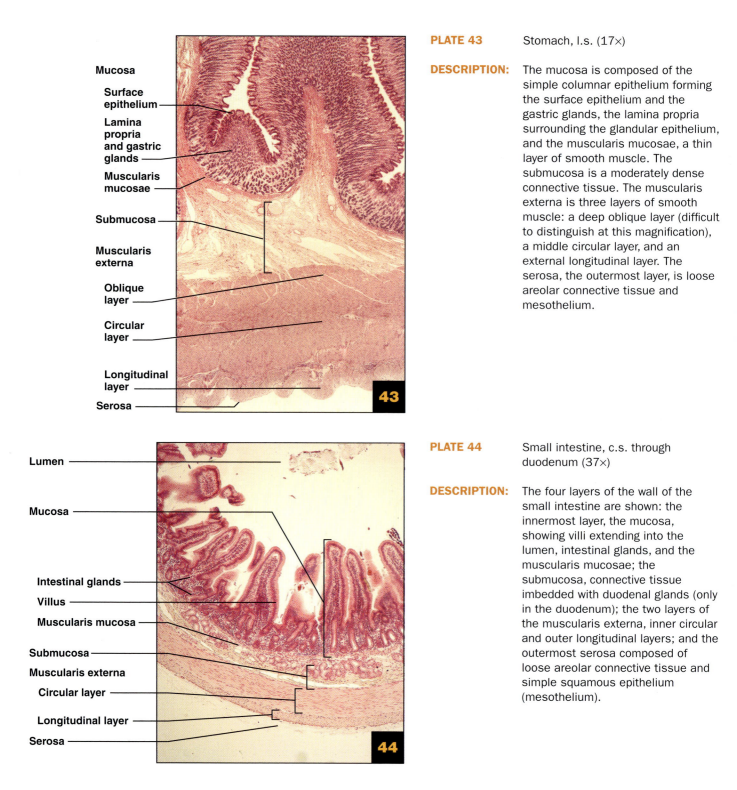

Mucosa
Surface epithelium
Lamina propria and gastric glands
Muscularis mucosae
Submucosa
Muscularis externa
Oblique layer
Circular layer
Longitudinal layer
Serosa

43

Lumen
Mucosa
Intestinal glands
Villus
Muscularis mucosa
Submucosa
Muscularis externa
Circular layer
Longitudinal layer
Serosa

44

PLATE 43   Stomach, l.s. (17×)

DESCRIPTION: The mucosa is composed of the simple columnar epithelium forming the surface epithelium and the gastric glands, the lamina propria surrounding the glandular epithelium, and the muscularis mucosae, a thin layer of smooth muscle. The submucosa is a moderately dense connective tissue. The muscularis externa is three layers of smooth muscle: a deep oblique layer (difficult to distinguish at this magnification), a middle circular layer, and an external longitudinal layer. The serosa, the outermost layer, is loose areolar connective tissue and mesothelium.

PLATE 44   Small intestine, c.s. through duodenum (37×)

DESCRIPTION: The four layers of the wall of the small intestine are shown: the innermost layer, the mucosa, showing villi extending into the lumen, intestinal glands, and the muscularis mucosae; the submucosa, connective tissue imbedded with duodenal glands (only in the duodenum); the two layers of the muscularis externa, inner circular and outer longitudinal layers; and the outermost serosa composed of loose areolar connective tissue and simple squamous epithelium (mesothelium).

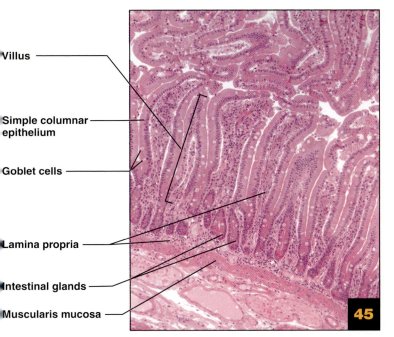

Villus

Simple columnar
epithelium

Goblet cells

Lamina propria

Intestinal glands

Muscularis mucosa

45

PLATE 45　Mucosal layer of small intestine, from the jejunum (76×)

DESCRIPTION:　Details of the mucosal layer are shown. Villi, lined with simple columnar epithelium and goblet cells, extend into the intestinal lumen; intestinal glands at the base of the villi produce digestive secretions; muscularis mucosae, a smooth muscle layer, forms the outermost layer of the mucosa. Refer to Plate 4 for high power view of mucosa. View the microvilli that make up the brush border on the apical surface of the columnar epithelial cells.

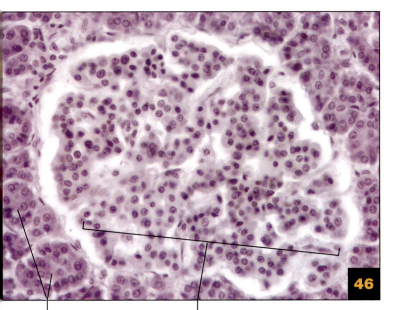

46

Exocrine pancreas,
acinar cells

Endocrine pancreas,
pancreatic islet

PLATE 46　Pancreas (350×)

DESCRIPTION:　The glandular cells in the center of the field make up the pancreatic islet, the endocrine portion of the pancreas that produces insulin (beta cells), glucagon (alpha cells), and somatostatin (delta cells). These hormones are secreted into capillaries surrounding these cells. The exocrine pancreas, the acinar cells surround the islet. These cell produce digestive secretions that empty into the duodenum via ducts.

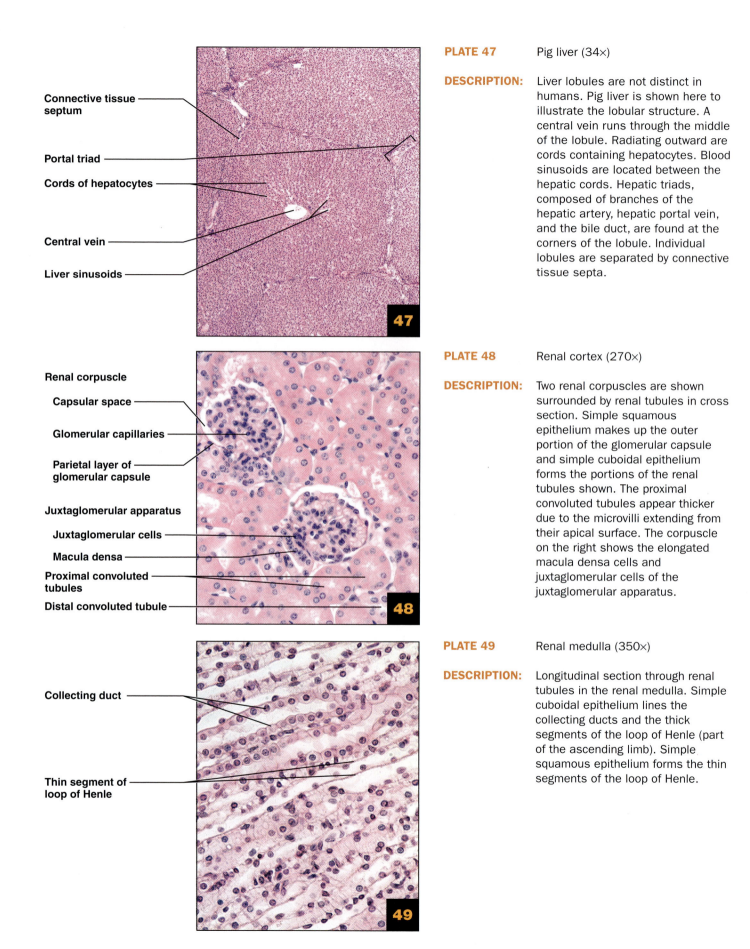

Connective tissue septum

Portal triad

Cords of hepatocytes

Central vein

Liver sinusoids

47

PLATE 47   Pig liver (34×)

DESCRIPTION: Liver lobules are not distinct in humans. Pig liver is shown here to illustrate the lobular structure. A central vein runs through the middle of the lobule. Radiating outward are cords containing hepatocytes. Blood sinusoids are located between the hepatic cords. Hepatic triads, composed of branches of the hepatic artery, hepatic portal vein, and the bile duct, are found at the corners of the lobule. Individual lobules are separated by connective tissue septa.

Renal corpuscle

    Capsular space

    Glomerular capillaries

    Parietal layer of glomerular capsule

Juxtaglomerular apparatus

    Juxtaglomerular cells

    Macula densa

Proximal convoluted tubules

Distal convoluted tubule

48

PLATE 48   Renal cortex (270×)

DESCRIPTION: Two renal corpuscles are shown surrounded by renal tubules in cross section. Simple squamous epithelium makes up the outer portion of the glomerular capsule and simple cuboidal epithelium forms the portions of the renal tubules shown. The proximal convoluted tubules appear thicker due to the microvilli extending from their apical surface. The corpuscle on the right shows the elongated macula densa cells and juxtaglomerular cells of the juxtaglomerular apparatus.

Collecting duct

Thin segment of loop of Henle

49

PLATE 49   Renal medulla (350×)

DESCRIPTION: Longitudinal section through renal tubules in the renal medulla. Simple cuboidal epithelium lines the collecting ducts and the thick segments of the loop of Henle (part of the ascending limb). Simple squamous epithelium forms the thin segments of the loop of Henle.

Lumen of the bladder

Mucosa

Transitional epithelium

Lamina propria

Detrusor muscle

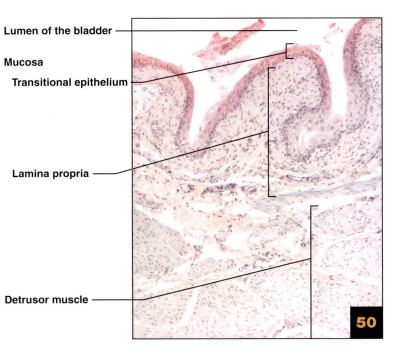

50

PLATE 50   Urinary bladder (97×)

DESCRIPTION: The transitional epithelium lining the urinary bladder and the lamina propria, composed of loose areolar connective tissue, together make up the mucosa of the bladder. The thick smooth muscle layer, the detrusor muscle, is made up of three indistinct layers. The outer layer, the adventitia, is not shown here.

Spermatogonia   Interstitial cells

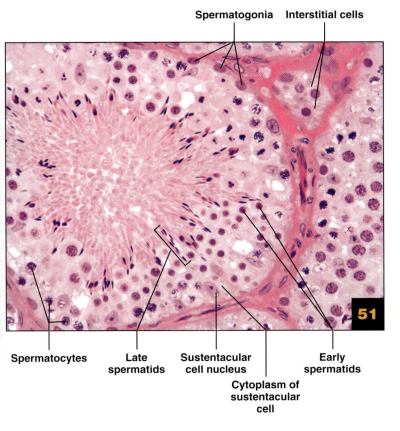

51

Spermatocytes   Late spermatids   Sustentacular cell nucleus   Early spermatids

Cytoplasm of sustentacular cell

PLATE 51   Testes (340×)

DESCRIPTION: This cross section through a seminferous tubule shows spermatogenic cells embedded within columnar sustentacular cells. Stem cells, spermatogonia, are located at the periphery of the tubule. As cells move through the tubule toward the lumen, sperm are formed. Clusters of cells located between the tubules, interstitial cells, secrete testosterone.

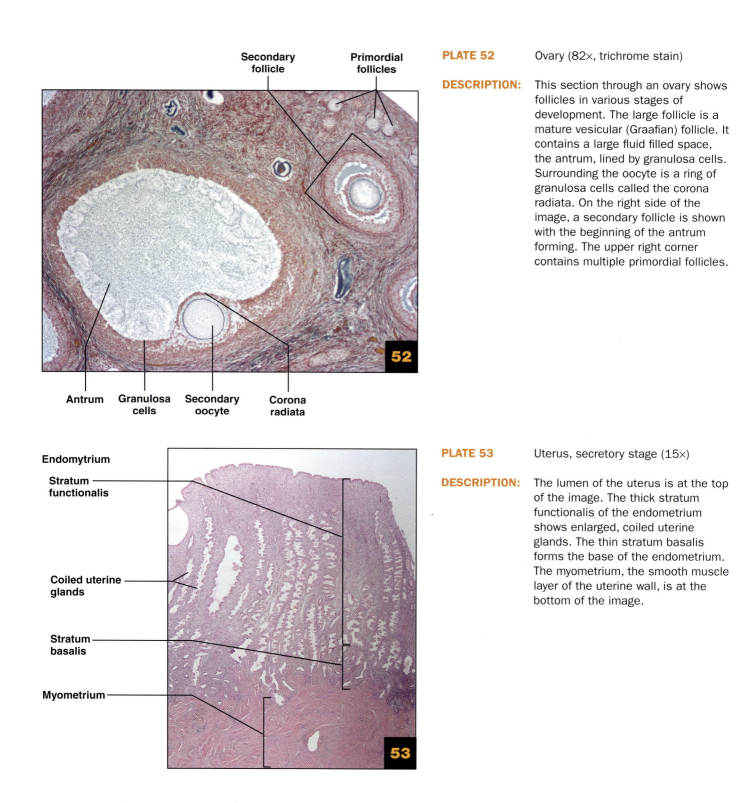

Secondary follicle

Primordial follicles

Antrum **Granulosa cells** **Secondary oocyte** **Corona radiata**

52

PLATE 52 Ovary (82×, trichrome stain)

DESCRIPTION: This section through an ovary shows follicles in various stages of development. The large follicle is a mature vesicular (Graafian) follicle. It contains a large fluid filled space, the antrum, lined by granulosa cells. Surrounding the oocyte is a ring of granulosa cells called the corona radiata. On the right side of the image, a secondary follicle is shown with the beginning of the antrum forming. The upper right corner contains multiple primordial follicles.

Endomytrium

Stratum functionalis

Coiled uterine glands

Stratum basalis

Myometrium

53

PLATE 53 Uterus, secretory stage (15×)

DESCRIPTION: The lumen of the uterus is at the top of the image. The thick stratum functionalis of the endometrium shows enlarged, coiled uterine glands. The thin stratum basalis forms the base of the endometrium. The myometrium, the smooth muscle layer of the uterine wall, is at the bottom of the image.

Colloid filled follicles

Follicle cells

PLATE 54   Thyroid gland (350×)

DESCRIPTION: The thyroid gland is composed of spherical follicles formed by simple cuboidal epithelial cells, follicle cells. The center of each follicle is filled with a gel-like substance called colloid that contains proteins needed for the formation of thyroid hormone. Thyroid hormone is produced by the follicle cells and secreted into the capillaries that surround the follicles. In this image the colloid has pulled away from the follicle cells, an artifact of slide preparation.

Capsule

Zona glomerulosa

Zona fasciculata

Zona reticularis

Medulla

PLATE 55   Adrenal gland, section (35×)

DESCRIPTION: This section through the adrenal gland shows the medulla, light pink oval in the bottom of the image, whose cells secrete epinephrine and norepinephrine; the portions of the cortex: the cells of the thin zona reticularis and the thicker zona fasciculata secrete glucocortidoid hormones, and the cells of the superficial zona glomerulosa secrete mineralocorticoid hormones. The connective tissue forming the adrenal capsule is at the top of the image.

Part II

BONES OF
THE HUMAN
SKELETON

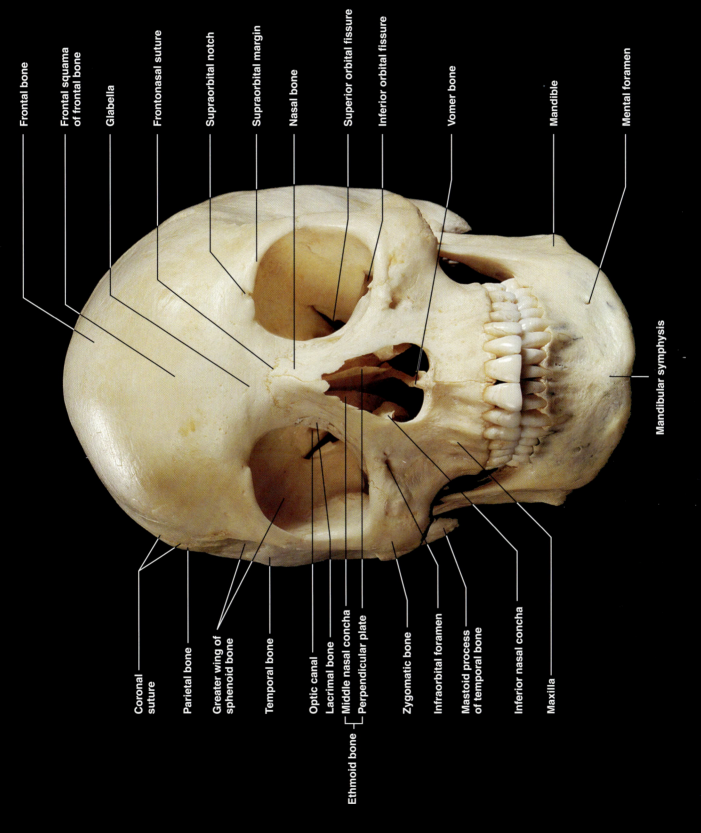

Frontal bone

Frontal squama of frontal bone

Glabella

Frontonasal suture

Supraorbital notch

Supraorbital margin

Nasal bone

Superior orbital fissure

Inferior orbital fissure

Vomer bone

Mandible

Mental foramen

Coronal suture

Parietal bone

Greater wing of sphenoid bone

Temporal bone

Optic canal

Lacrimal bone

Ethmoid bone — Middle nasal concha

Perpendicular plate

Zygomatic bone

Infraorbital foramen

Mastoid process of temporal bone

Inferior nasal concha

Maxilla

Mandibular symphysis

Figure 1 Skull, anterior view.

27

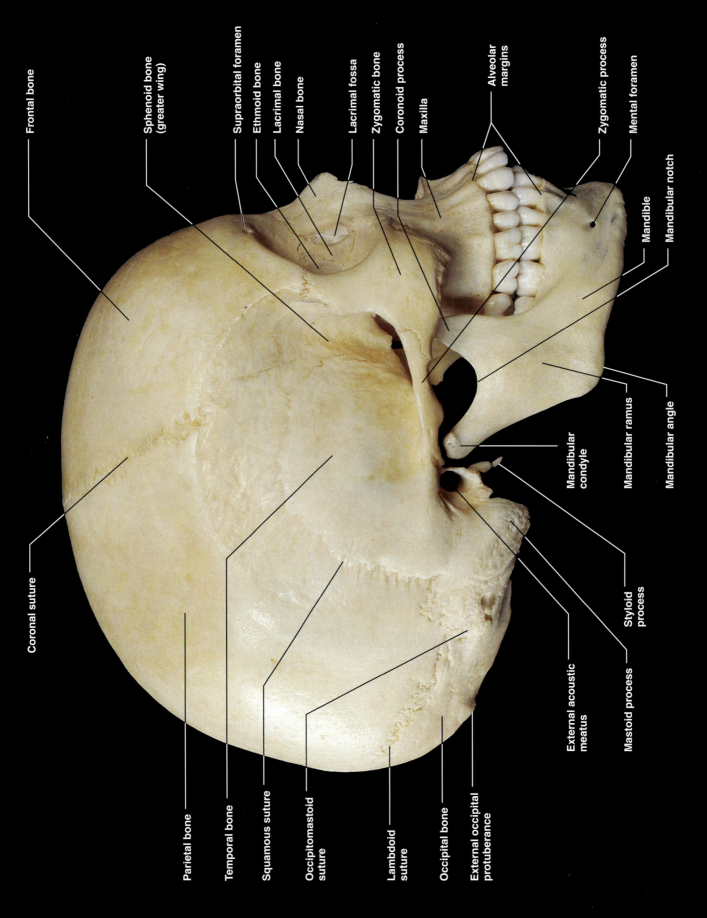

Figure 2 Skull, right external view of lateral surface.

Frontal bone

Sphenoid bone (greater wing)

Supraorbital foramen

Ethmoid bone

Lacrimal bone

Nasal bone

Lacrimal fossa

Zygomatic bone

Coronoid process

Maxilla

Alveolar margins

Zygomatic process

Mental foramen

Mandible

Mandibular notch

Mandibular condyle

Mandibular ramus

Mandibular angle

External acoustic meatus

Mastoid process

Styloid process

Coronal suture

Parietal bone

Temporal bone

Squamous suture

Occipitomastoid suture

Lambdoid suture

Occipital bone

External occipital protuberance

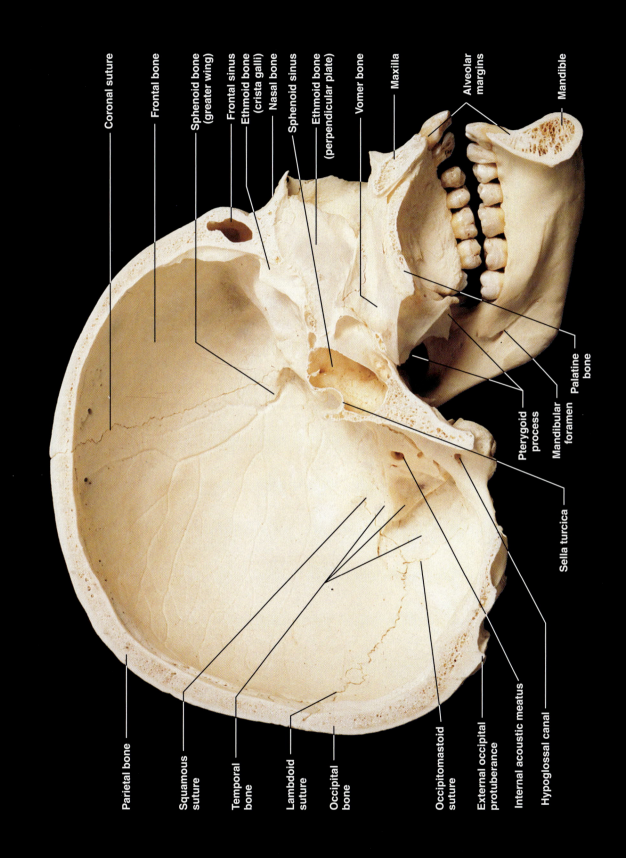

Coronal suture

Frontal bone

Sphenoid bone
(greater wing)

Frontal sinus

Ethmoid bone
(crista galli)

Nasal bone

Sphenoid sinus

Ethmoid bone
(perpendicular plate)

Vomer bone

Maxilla

Alveolar
margins

Mandible

Parietal bone

Squamous
suture

Temporal
bone

Lambdoid
suture

Occipital
bone

Occipitomastoid
suture

External occipital
protuberance

Internal acoustic meatus

Hypoglossal canal

Sella turcica

Pterygoid
process

Mandibular
foramen

Palatine
bone

Figure 3 Skull, internal view of left lateral aspect.

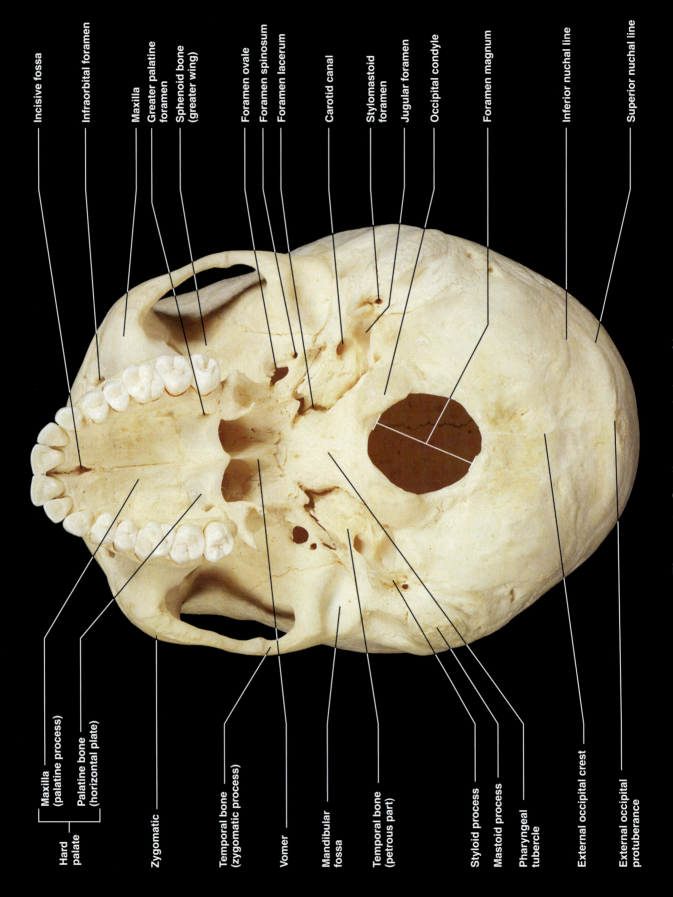

Incisive fossa

Infraorbital foramen

Maxilla

Greater palatine foramen

Sphenoid bone (greater wing)

Foramen ovale

Foramen spinosum

Foramen lacerum

Carotid canal

Stylomastoid foramen

Jugular foramen

Occipital condyle

Foramen magnum

Inferior nuchal line

Superior nuchal line

Maxilla (palatine process)

Palatine bone (horizontal plate)

Hard palate

Zygomatic

Temporal bone (zygomatic process)

Vomer

Mandibular fossa

Temporal bone (petrous part)

Styloid process

Mastoid process

Pharyngeal tubercle

External occipital crest

External occipital protuberance

Figure 4 Skull, external view of base.

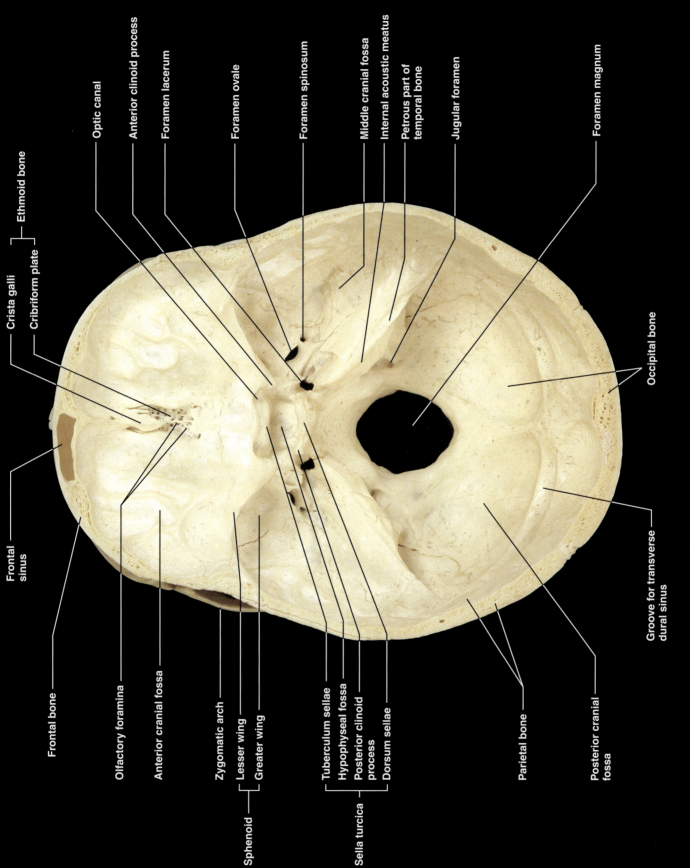

Crista galli ⎤
Ethmoid bone
Cribriform plate ⎦

Optic canal

Anterior clinoid process

Foramen lacerum

Foramen ovale

Foramen spinosum

Middle cranial fossa

Internal acoustic meatus

Petrous part of temporal bone

Jugular foramen

Foramen magnum

Frontal sinus

Occipital bone

Frontal bone

Olfactory foramina

Anterior cranial fossa

Zygomatic arch

Lesser wing ⎤
Sphenoid
Greater wing ⎦

Tuberculum sellae ⎤
Hypophyseal fossa
Sella turcica
Posterior clinoid process
Dorsum sellae ⎦

Parietal bone

Posterior cranial fossa

Groove for transverse dural sinus

Figure 5 Skull, internal view of base.

31

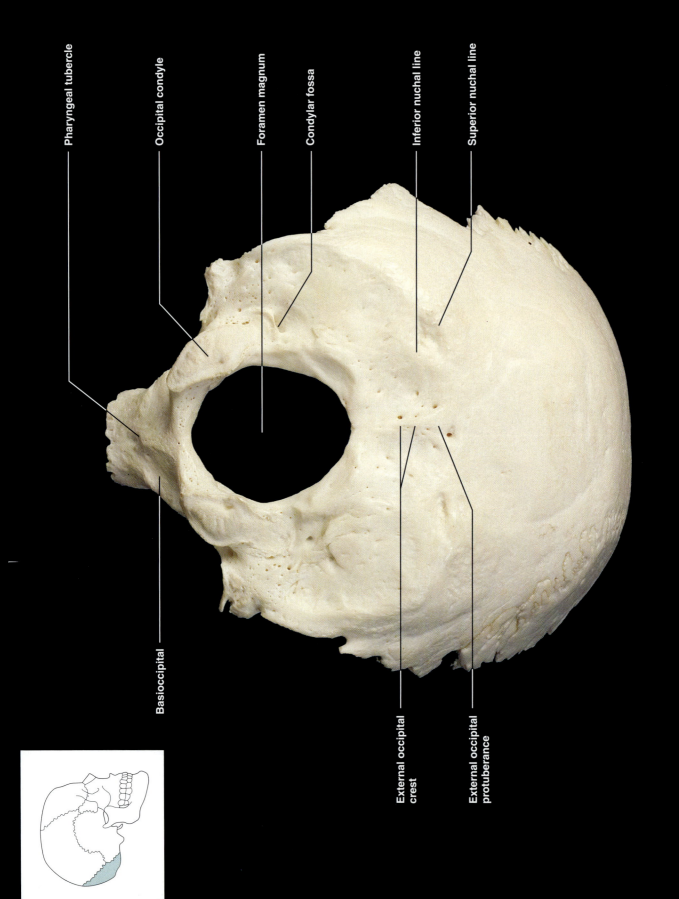

Pharyngeal tubercle

Occipital condyle

Foramen magnum

Condylar fossa

Inferior nuchal line

Superior nuchal line

Basioccipital

External occipital crest

External occipital protuberance

Figure 6 Occipital bone, inferior external view.

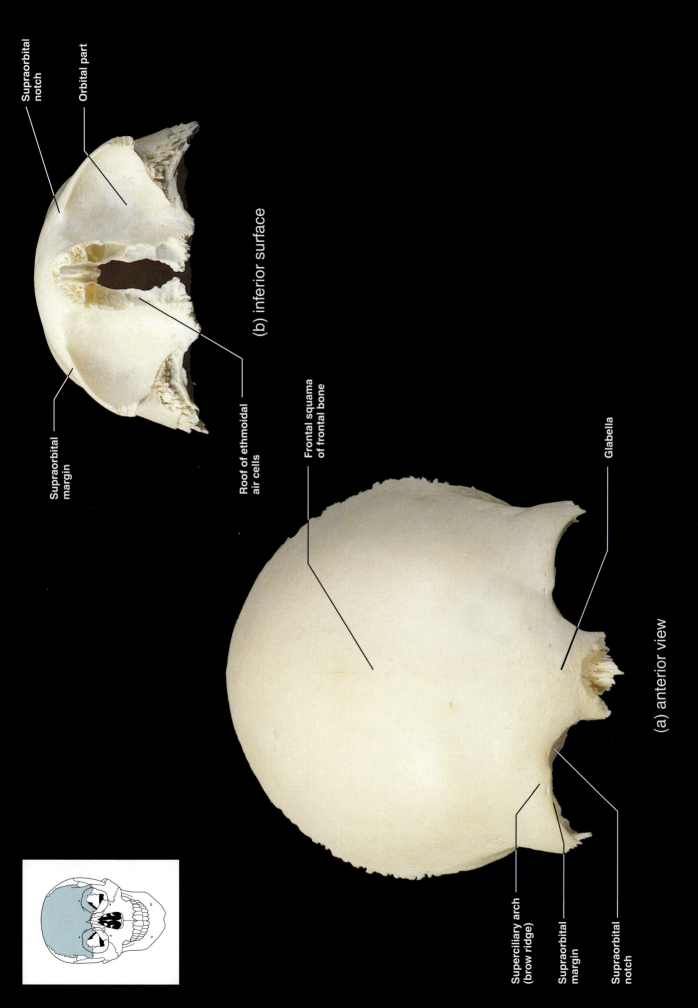

Supraorbital
notch

Orbital part

Supraorbital
margin

Roof of ethmoidal
air cells

(b) inferior surface

Frontal squama
of frontal bone

Glabella

Superciliary arch
(brow ridge)

Supraorbital
margin

Supraorbital
notch

(a) anterior view

Figure 7 Frontal bone.

Mastoid part

External acoustic meatus

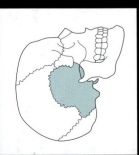

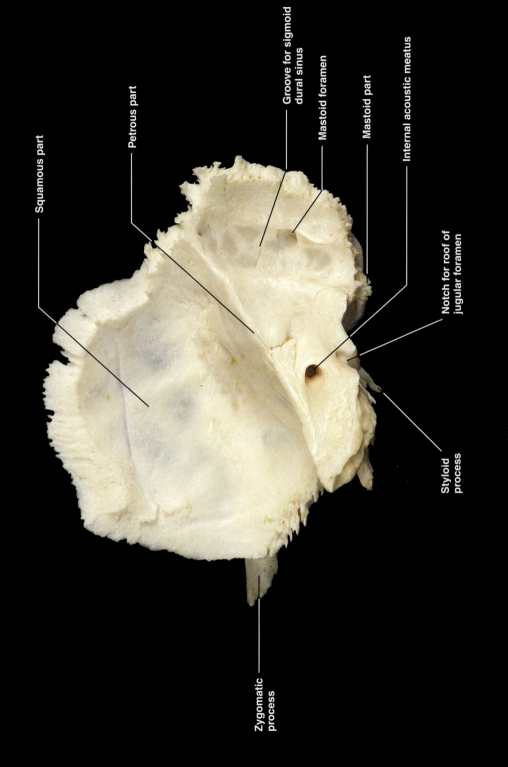

Squamous part

Petrous part

Groove for sigmoid dural sinus

Mastoid foramen

Mastoid part

Internal acoustic meatus

Notch for roof of jugular foramen

Styloid process

Zygomatic process

(b) right medial view

Figure 8 Temporal bone.

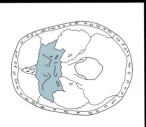

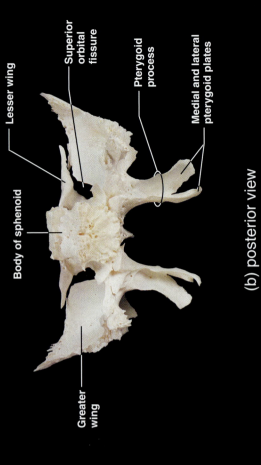

Lesser wing

Superior orbital fissure

Pterygoid process

Medial and lateral pterygoid plates

Body of sphenoid

Greater wing

(b) posterior view

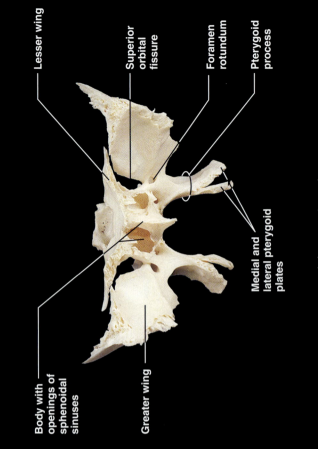

Lesser wing

Superior orbital fissure

Foramen rotundum

Pterygoid process

Body with openings of sphenoidal sinuses

Medial and lateral pterygoid plates

Greater wing

(c) anterior view

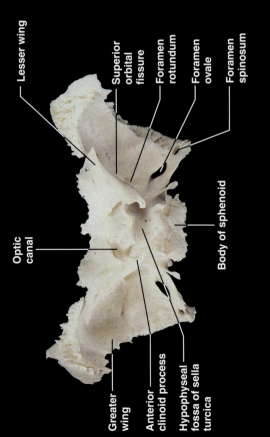

Lesser wing

Superior orbital fissure

Foramen rotundum

Foramen ovale

Foramen spinosum

Optic canal

Body of sphenoid

Greater wing

Anterior clinoid process

Hypophyseal fossa of sella turcica

(a) superior view

Figure 9 Sphenoid bone.

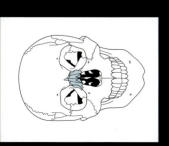

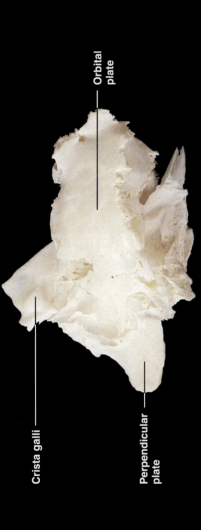

Crista galli

Perpendicular plate

Orbital plate

(a) left lateral surface

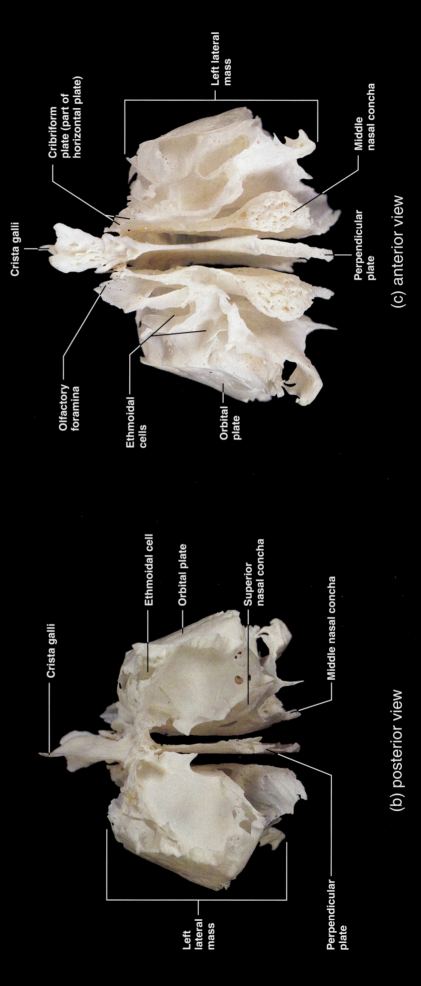

Crista galli

Cribriform plate (part of horizontal plate)

Left lateral mass

Olfactory foramina

Ethmoidal cells

Orbital plate

Middle nasal concha

Perpendicular plate

(c) anterior view

Crista galli

Ethmoidal cell

Orbital plate

Superior nasal concha

Middle nasal concha

Left lateral mass

Perpendicular plate

(b) posterior view

Figure 10 Ethmoid bone.

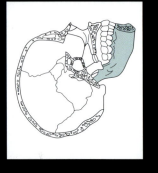

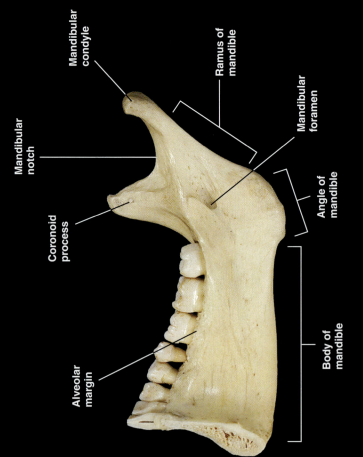

Mandibular
condyle

Ramus of
mandible

Mandibular
notch

Mandibular
foramen

Coronoid
process

Angle of
mandible

Alveolar
margin

Body of
mandible

(b) right medial view

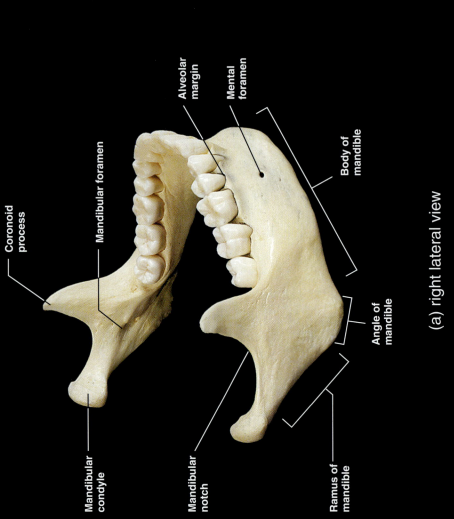

Coronoid
process

Mandibular foramen

Mandibular
condyle

Mandibular
notch

Ramus of
mandible

Alveolar
margin

Mental
foramen

Body of
mandible

Angle of
mandible

(a) right lateral view

Figure 11 Mandible.

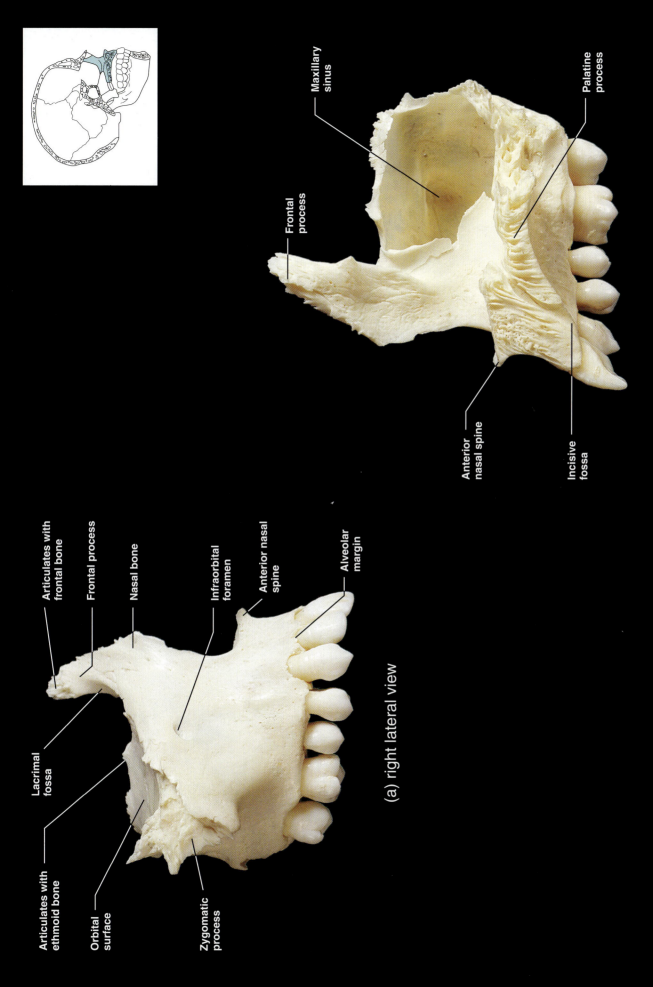

Articulates with
ethmoid bone

Frontal process

Nasal bone

Infraorbital
foramen

Anterior nasal
spine

Alveolar
margin

Lacrimal
fossa

Articulates with
frontal bone

Orbital
surface

Zygomatic
process

(a) right lateral view

Frontal
process

Maxillary
sinus

Anterior
nasal spine

Incisive
fossa

Palatine
process

(b) right medial view

Figure 12 Maxilla.

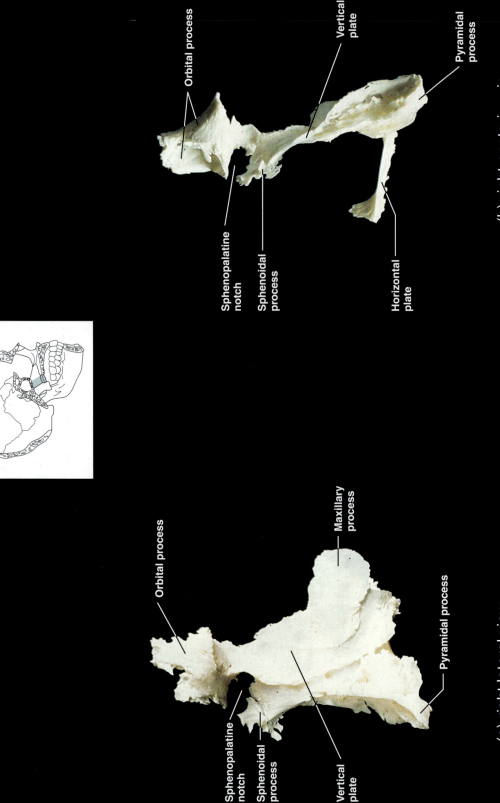

Orbital process

Sphenopalatine notch

Sphenoidal process

Vertical plate

Maxillary process

Pyramidal process

(a) right lateral view

Orbital process

Vertical plate

Pyramidal process

Sphenopalatine notch

Sphenoidal process

Horizontal plate

(b) right posterior view

Supraorbital notch

Optic canal

Roof of orbit

Orbital plate of
frontal bone

Lesser wing of
sphenoid bone

Lateral wall of orbit

Zygomatic process
of frontal bone

Greater wing of
sphenoid bone

Orbital surface of
zygomatic bone

Superior orbital
fissure

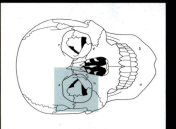

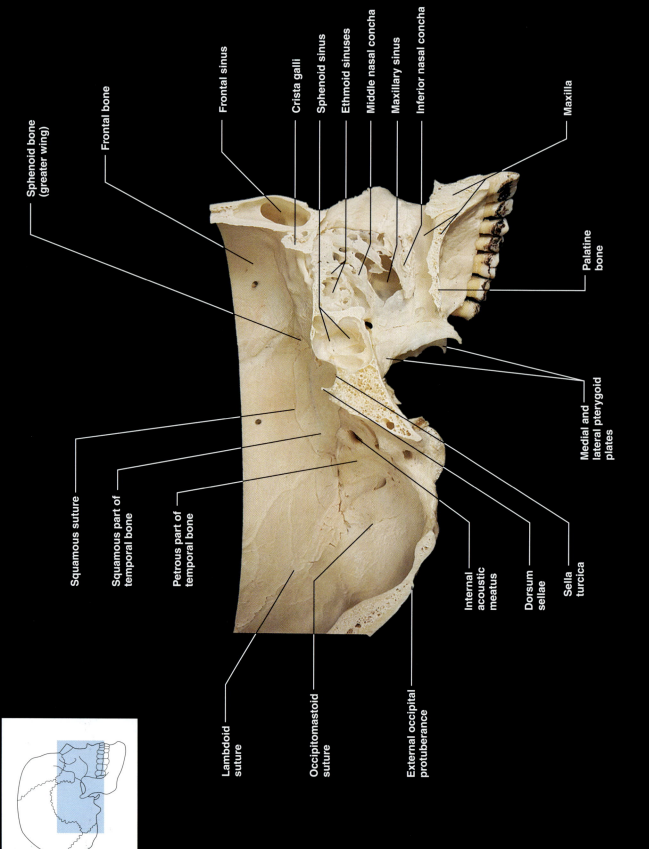

Figure 15 Nasal cavity, left lateral wall.

Sphenoid bone (greater wing)

Frontal bone

Frontal sinus

Crista galli

Sphenoid sinus

Ethmoid sinuses

Middle nasal concha

Maxillary sinus

Inferior nasal concha

Maxilla

Palatine bone

Medial and lateral pterygoid plates

Squamous suture

Squamous part of temporal bone

Petrous part of temporal bone

Internal acoustic meatus

Dorsum sellae

Sella turcica

Lambdoid suture

Occipitomastoid suture

External occipital protuberance

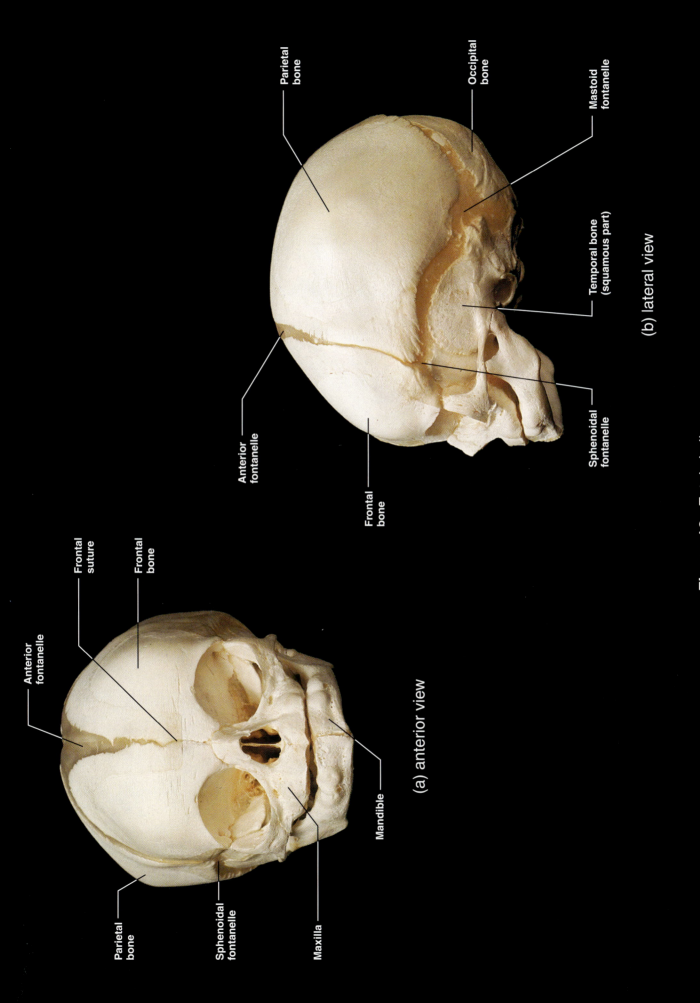

Anterior fontanelle

Frontal suture

Frontal bone

Parietal bone

Sphenoidal fontanelle

Maxilla

Mandible

(a) anterior view

Parietal bone

Anterior fontanelle

Occipital bone

Mastoid fontanelle

Temporal bone (squamous part)

Frontal bone

Sphenoidal fontanelle

(b) lateral view

Figure 16 Fetal skull.

44

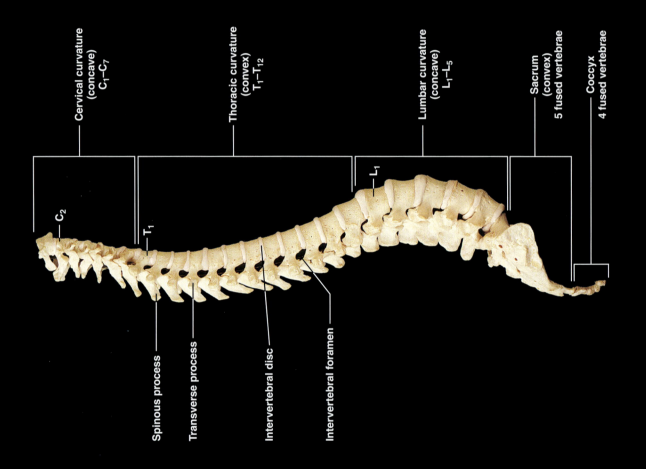

Cervical curvature
(concave)
C_1–C_7

Thoracic curvature
(convex)
T_1–T_{12}

Lumbar curvature
(concave)
L_1–L_5

Sacrum
(convex)
5 fused vertebrae

Coccyx
4 fused vertebrae

C_2

T_1

L_1

Spinous process

Transverse process

Intervertebral disc

Intervertebral foramen

(a) right lateral view

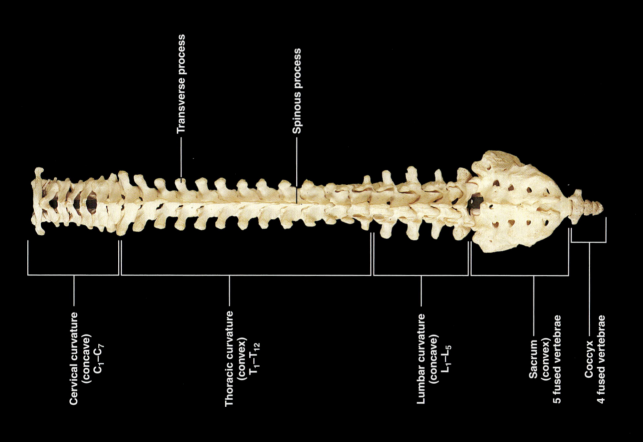

Transverse process

Spinous process

Cervical curvature
(concave)
C_1-C_7

Thoracic curvature
(convex)
T_1-T_{12}

Lumbar curvature
(concave)
L_1-L_5

Sacrum
(convex)
5 fused vertebrae

Coccyx
4 fused vertebrae

(b) posterior view

Figure 17 Articulated vertebral column.

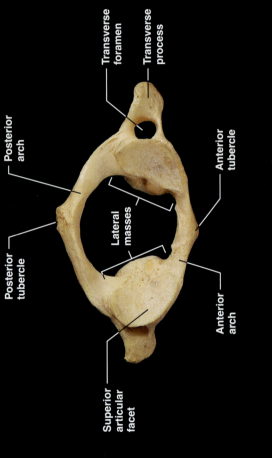

Posterior
arch

Transverse
foramen

Transverse
process

Posterior
tubercle

Lateral
masses

Anterior
tubercle

Superior
articular
facet

Anterior
arch

(a) atlas, superior view

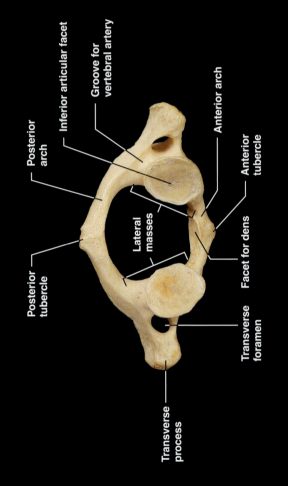

Posterior
arch

Inferior articular facet

Groove for
vertebral artery

Anterior arch

Posterior
tubercle

Lateral
masses

Anterior
tubercle

Transverse
process

Facet for dens

Transverse
foramen

(b) atlas, inferior view

C₁
C₂

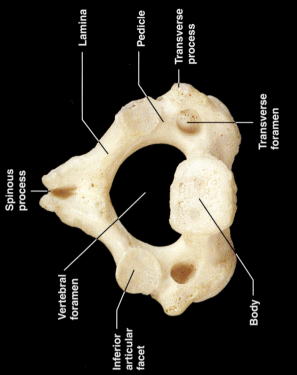

Spinous
process

Lamina

Pedicle

Transverse
process

Vertebral
foramen

Inferior
articular
facet

Transverse
foramen

Body

(d) axis, inferior view

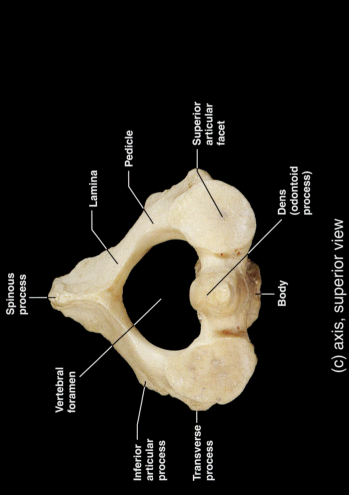

Spinous
process

Lamina

Pedicle

Superior
articular
facet

Vertebral
foramen

Inferior
articular
process

Dens
(odontoid
process)

Body

Transverse
process

(c) axis, superior view

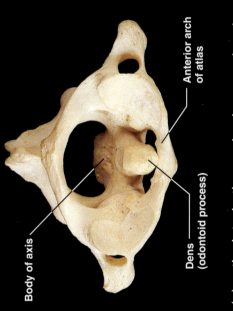

Body of axis

Anterior arch
of atlas

Dens
(odontoid process)

(e) articulated atlas and axis, superior view

Figure 18 Various views of vertebrae C_1 and C_2.

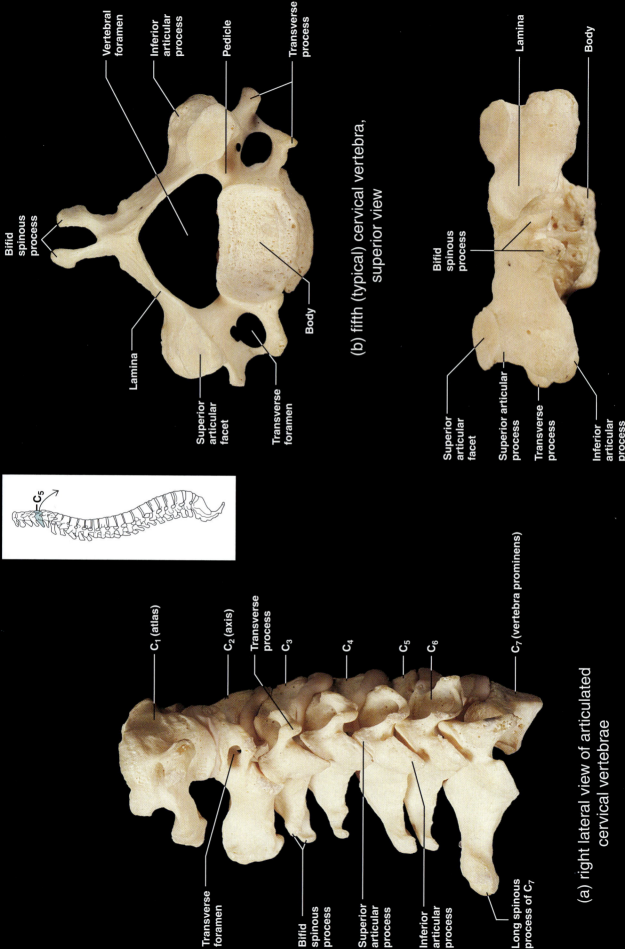

Vertebral foramen

Inferior articular process

Pedicle

Transverse process

Bifid spinous process

Lamina

Superior articular facet

Transverse foramen

Body

(b) fifth (typical) cervical vertebra, superior view

Lamina

Body

Bifid spinous process

Superior articular facet

Superior articular process

Transverse process

Inferior articular process

C₁ (atlas)

C₂ (axis)

Transverse process

C₃

C₄

C₅

C₆

C₇ (vertebra prominens)

Transverse foramen

Bifid spinous process

Superior articular process

Inferior articular process

Long spinous process of C₇

(a) right lateral view of articulated cervical vertebrae

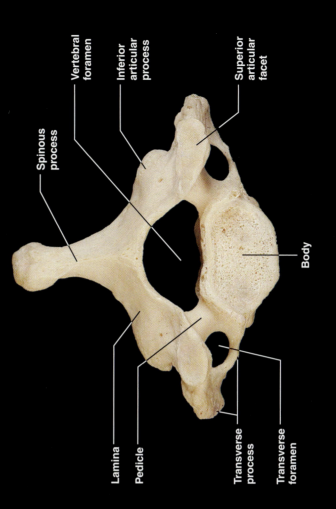

Spinous
process

Vertebral
foramen

Inferior
articular
process

Superior
articular
facet

Lamina

Pedicle

Transverse
process

Transverse
foramen

Body

(e) vertebra prominens (C₇), superior view

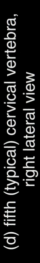

C₇

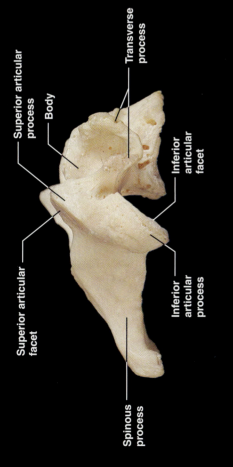

Superior articular
process

Body

Transverse
process

Inferior
articular
facet

Superior articular
facet

Inferior
articular
process

Spinous
process

(d) fifth (typical) cervical vertebra,
right lateral view

Figure 19 Cervical vertebrae.

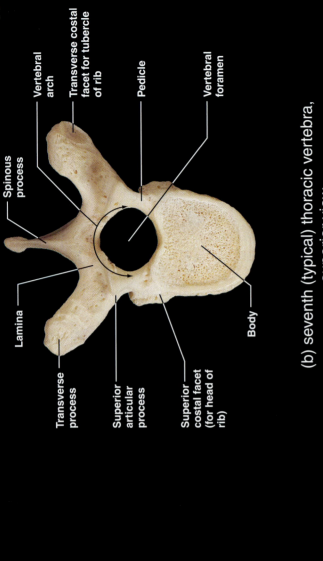

Spinous process

Lamina

Transverse process

Superior articular process

Superior costal facet (for head of rib)

Body

Vertebral arch

Transverse costal facet for tubercle of rib

Pedicle

Vertebral foramen

(b) seventh (typical) thoracic vertebra, superior view

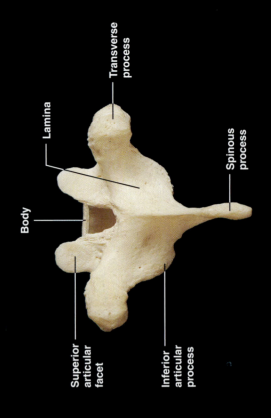

Body

Lamina

Transverse process

Spinous process

Superior articular facet

Inferior articular process

(c) seventh (typical) thoracic vertebra, posterior view

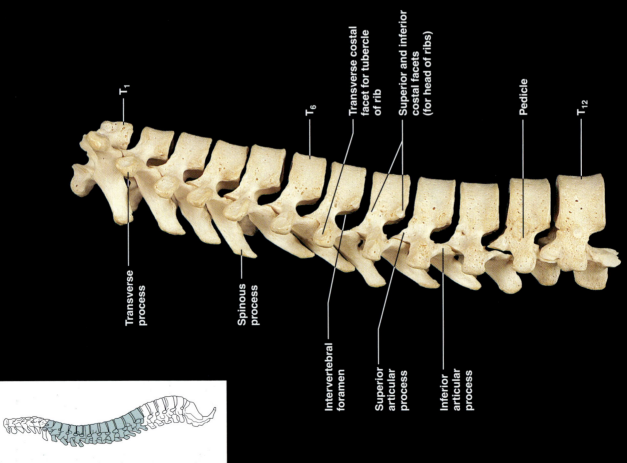

T₁

T₆

T₁₂

Transverse process

Spinous process

Intervertebral foramen

Transverse costal facet for tubercle of rib

Superior and inferior costal facets (for head of ribs)

Superior articular process

Inferior articular process

Pedicle

(a) articulated thoracic vertebrae, right lateral view

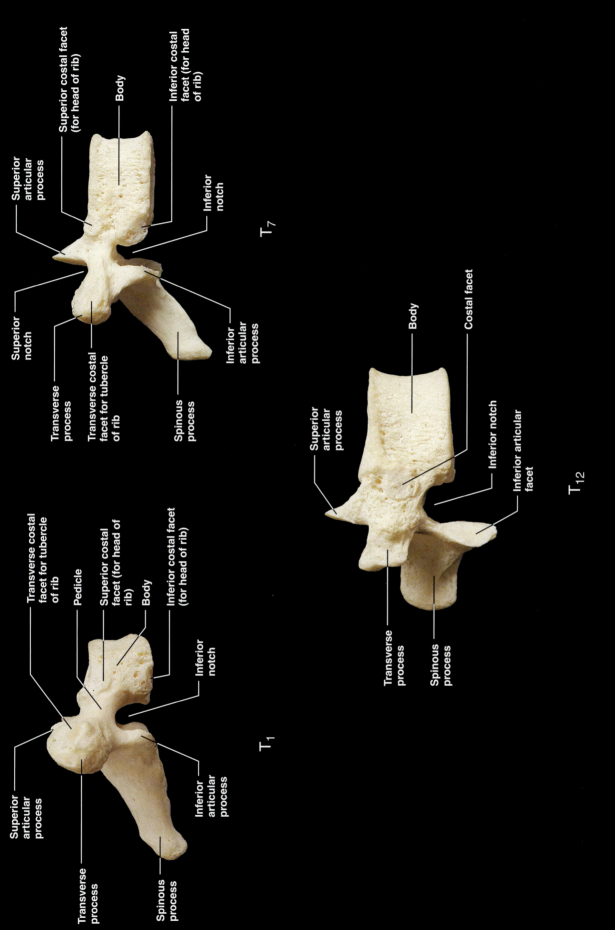

T7

Superior costal facet (for head of rib)

Body

Inferior costal facet (for head of rib)

Superior articular process

Inferior notch

Superior notch

Transverse process

Transverse costal facet for tubercle of rib

Spinous process

Inferior articular process

T1

Transverse costal facet for tubercle of rib

Pedicle

Superior costal facet (for head of rib)

Body

Inferior costal facet (for head of rib)

Inferior notch

Superior articular process

Transverse process

Spinous process

Inferior articular process

T12

Body

Costal facet

Superior articular process

Inferior notch

Inferior articular facet

Transverse process

Spinous process

(d) comparison of T$_1$, T$_7$, and T$_{12}$ in right lateral views

Figure 20 Thoracic vertebrae.

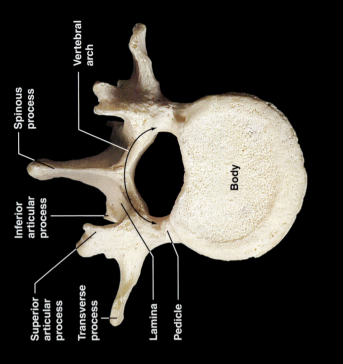

Spinous process

Vertebral arch

Inferior articular process

Superior articular process

Transverse process

Lamina

Pedicle

Body

(b) second lumbar vertebra, superior view

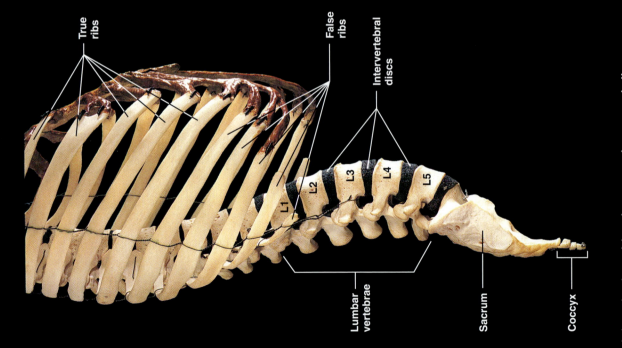

True ribs

False ribs

Intervertebral discs

L1

L2

L3

L4

L5

Lumbar vertebrae

Sacrum

Coccyx

(a) articulated lumbar vertebrae and rib cage, right lateral view

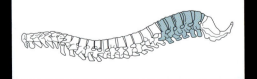

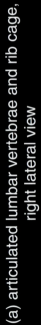

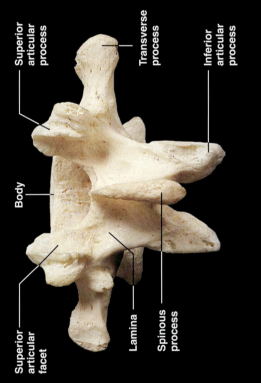

Superior
articular
process

Transverse
process

Inferior
articular
process

Body

Superior
articular
facet

Lamina

Spinous
process

(c) second lumbar vertebra, posterior view

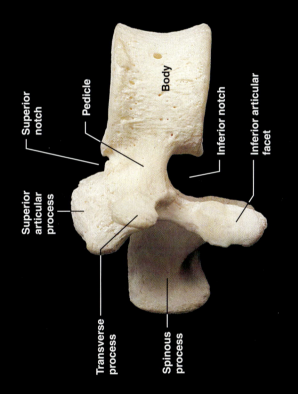

Superior
articular
process

Pedicle

Body

Superior
notch

Inferior notch

Inferior articular
facet

Transverse
process

Spinous
process

(d) second lumbar vertebra, right lateral view

Figure 21 Lumbar vertebrae.

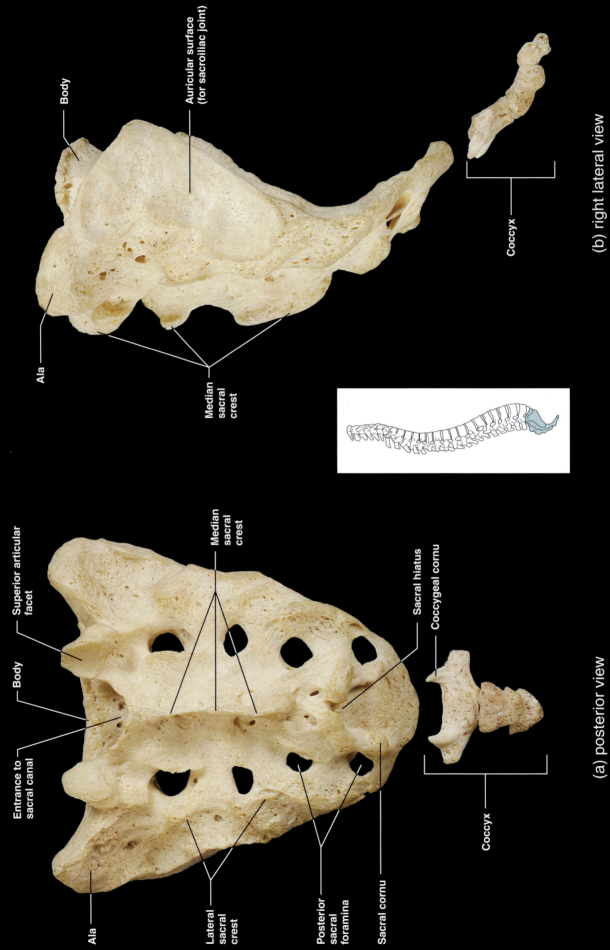

Body

Auricular surface
(for sacroiliac joint)

Ala

Median sacral
crest

Coccyx

(b) right lateral view

Superior articular
facet

Median sacral
crest

Body

Entrance to
sacral canal

Sacral hiatus

Coccygeal cornu

Ala

Lateral
sacral
crest

Posterior
sacral
foramina

Sacral cornu

Coccyx

(a) posterior view

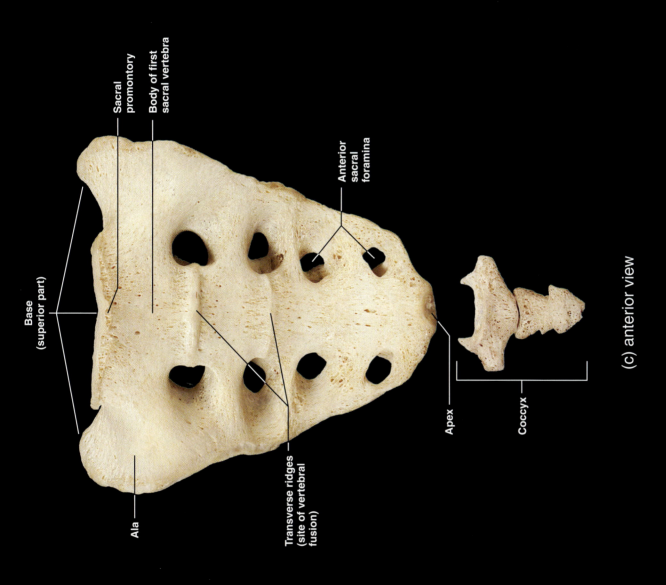

Base
(superior part)

Sacral
promontory

Body of first
sacral vertebra

Anterior
sacral
foramina

Ala

Transverse ridges
(site of vertebral
fusion)

Apex

Coccyx

(c) anterior view

Figure 22 Sacrum and coccyx.

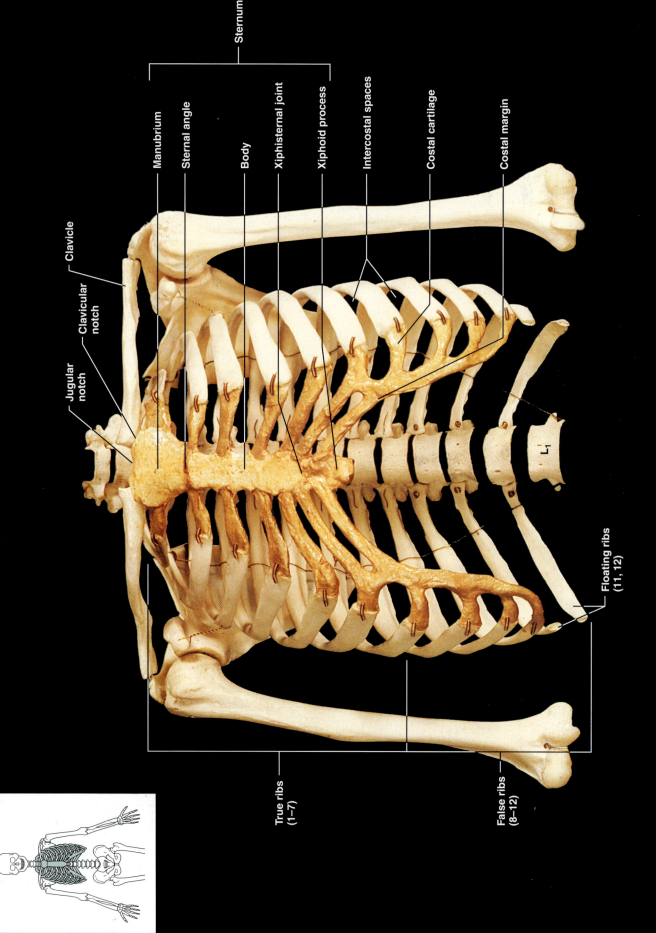

Sternum

Manubrium

Sternal angle

Body

Xiphisternal joint

Xiphoid process

Intercostal spaces

Costal cartilage

Costal margin

Clavicle

Clavicular notch

Jugular notch

True ribs (1–7)

False ribs (8–12)

Floating ribs (11, 12)

L_1

(a) anterior view

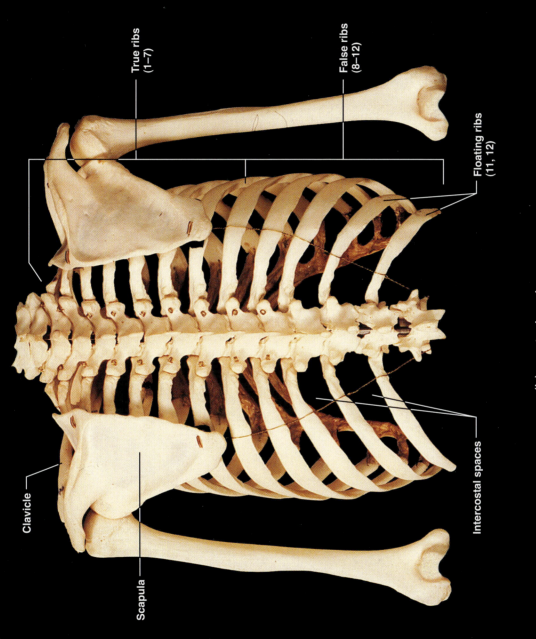

True ribs
(1–7)

False ribs
(8–12)

Floating ribs
(11, 12)

Clavicle

Scapula

Intercostal spaces

(b) posterior view

Figure 23 Thoracic cage

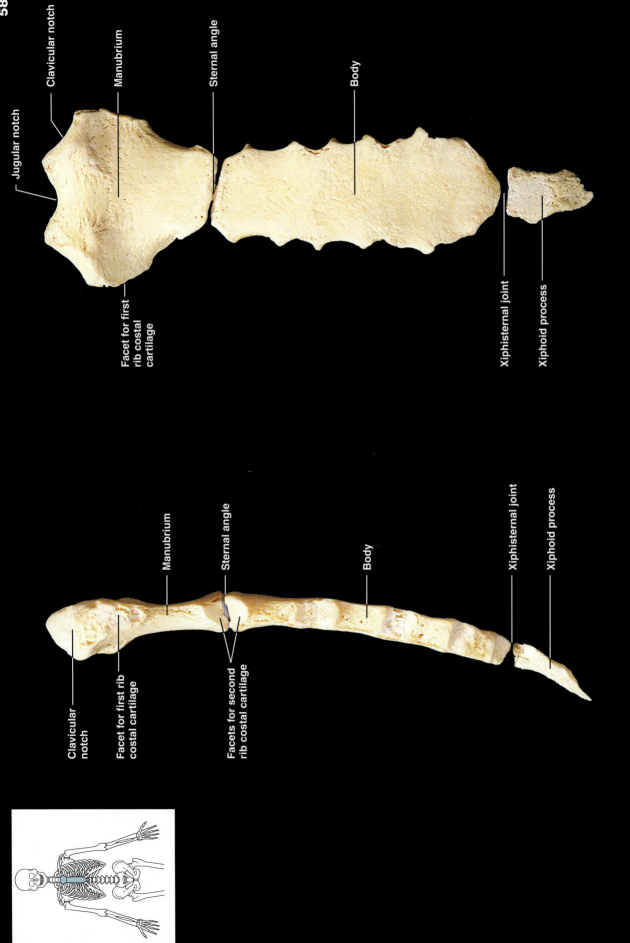

Jugular notch

Clavicular notch

Manubrium

Sternal angle

Body

Facet for first
rib costal
cartilage

Xiphisternal joint

Xiphoid process

(d) sternum, anterior view

Clavicular
notch

Manubrium

Sternal angle

Body

Facet for first rib
costal cartilage

Facets for second
rib costal cartilage

Xiphisternal joint

Xiphoid process

(c) sternum, right lateral view

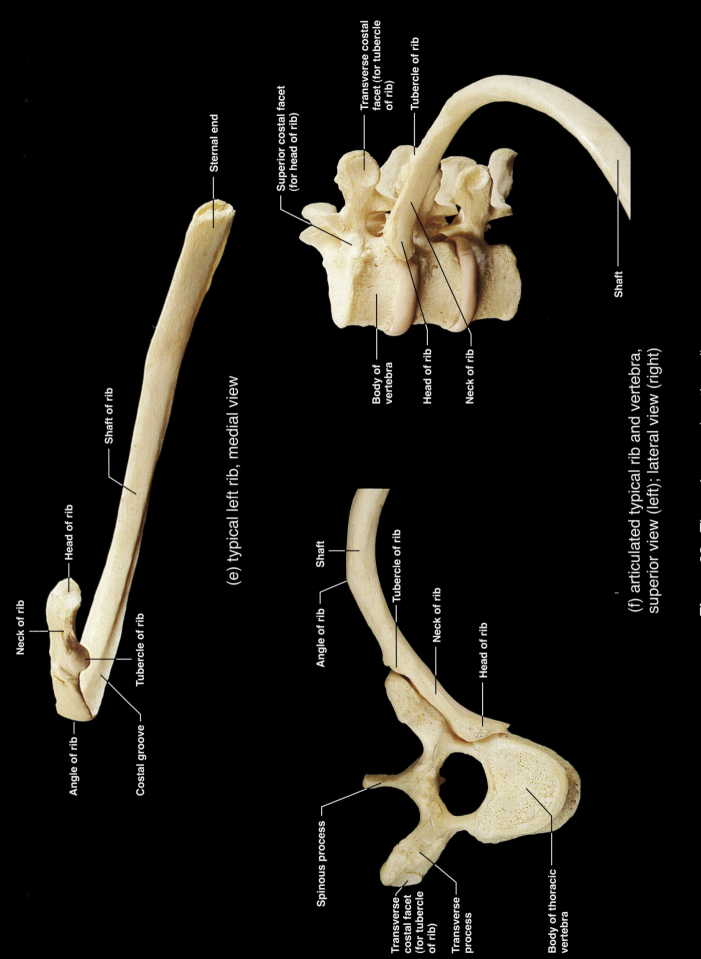

Neck of rib

Head of rib

Shaft of rib

Angle of rib

Tubercle of rib

Costal groove

Sternal end

(e) typical left rib, medial view

Superior costal facet (for head of rib)

Transverse costal facet (for tubercle of rib)

Tubercle of rib

Body of vertebra

Head of rib

Neck of rib

Shaft

Angle of rib

Shaft

Tubercle of rib

Neck of rib

Head of rib

Spinous process

Transverse costal facet (for tubercle of rib)

Transverse process

Body of thoracic vertebra

(f) articulated typical rib and vertebra, superior view (left); lateral view (right)

Figure 23 Thoracic cage (continued).

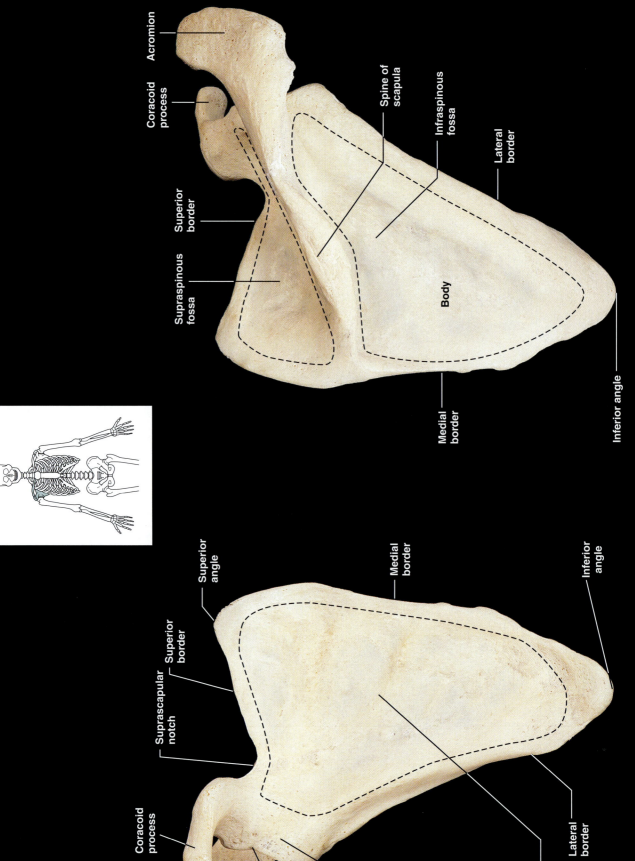

Acromion
Coracoid process
Superior border
Supraspinous fossa
Spine of scapula
Infraspinous fossa
Lateral border
Medial border
Inferior angle
Body

(b) right scapula, posterior view

Superior angle
Superior border
Suprascapular notch
Medial border
Inferior angle
Coracoid process
Acromion
Glenoid cavity
Lateral angle
Subscapular fossa
Lateral border

(a) right scapula, anterior view

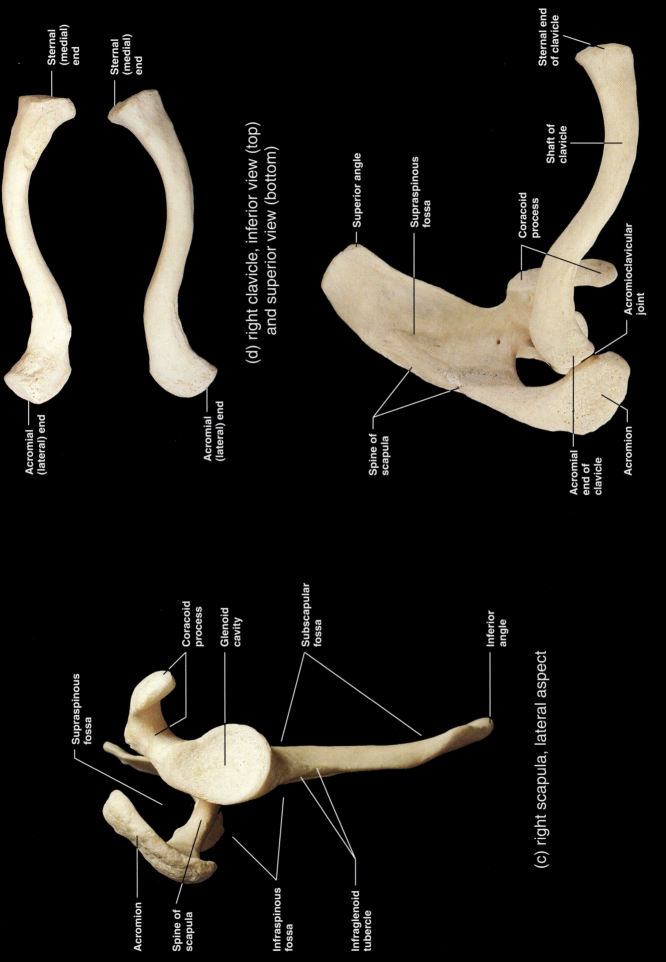

Sternal (medial) end

Sternal (medial) end

Acromial (lateral) end

Acromial (lateral) end

(d) right clavicle, inferior view (top) and superior view (bottom)

Superior angle

Supraspinous fossa

Coracoid process

Acromioclavicular joint

Sternal end of clavicle

Shaft of clavicle

Spine of scapula

Acromial end of clavicle

Acromion

(e) articulated right clavicle and scapula, superior view

Supraspinous fossa

Coracoid process

Glenoid cavity

Subscapular fossa

Inferior angle

Acromion

Spine of scapula

Infraspinous fossa

Infraglenoid tubercle

(c) right scapula, lateral aspect

Figure 24 Scapula and clavicle.

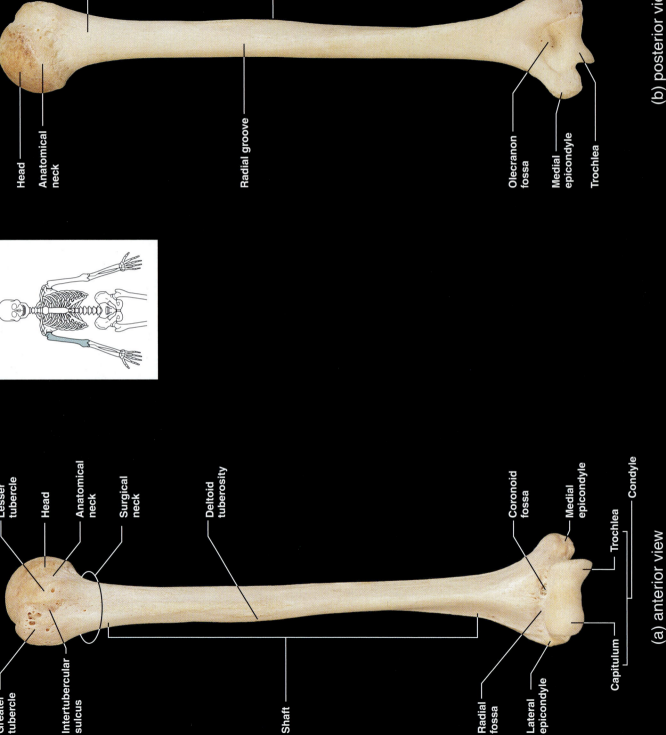

Head

Anatomical neck

Radial groove

Olecranon fossa

Medial epicondyle

Trochlea

(b) posterior view

Lesser tubercle

Head

Anatomical neck

Surgical neck

Deltoid tuberosity

Coronoid fossa

Medial epicondyle

Trochlea

Condyle

Greater tubercle

Intertubercular sulcus

Shaft

Radial fossa

Lateral epicondyle

Capitulum

(a) anterior view

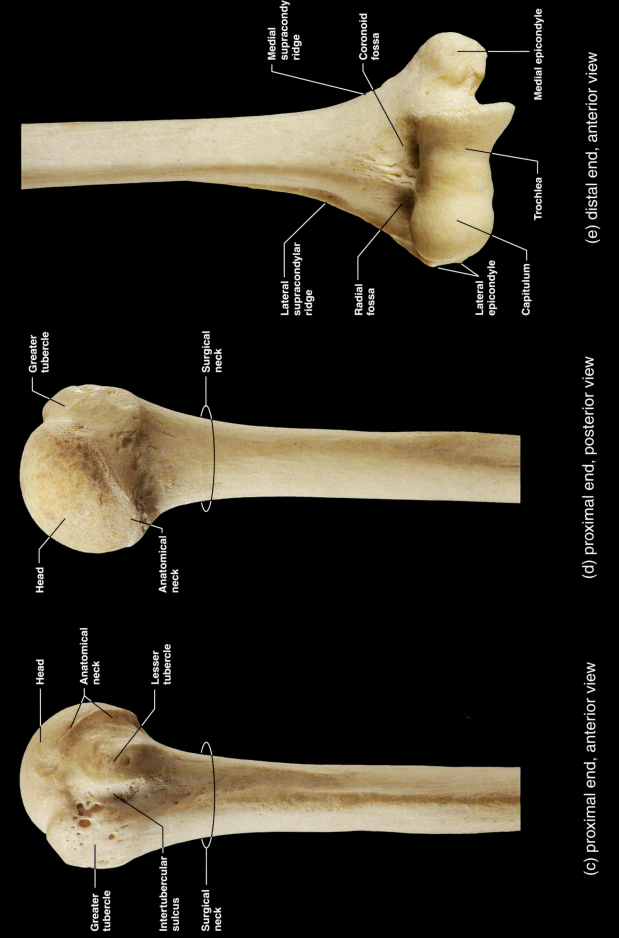

Medial supracondyl[ar] ridge

Coronoid fossa

Medial epicondyle

Lateral supracondylar ridge

Radial fossa

Lateral epicondyle

Trochlea

Capitulum

(e) distal end, anterior view

Greater tubercle

Surgical neck

Head

Anatomical neck

(d) proximal end, posterior view

Head

Anatomical neck

Lesser tubercle

Greater tubercle

Intertubercular sulcus

Surgical neck

(c) proximal end, anterior view

Figure 25 Right humerus.

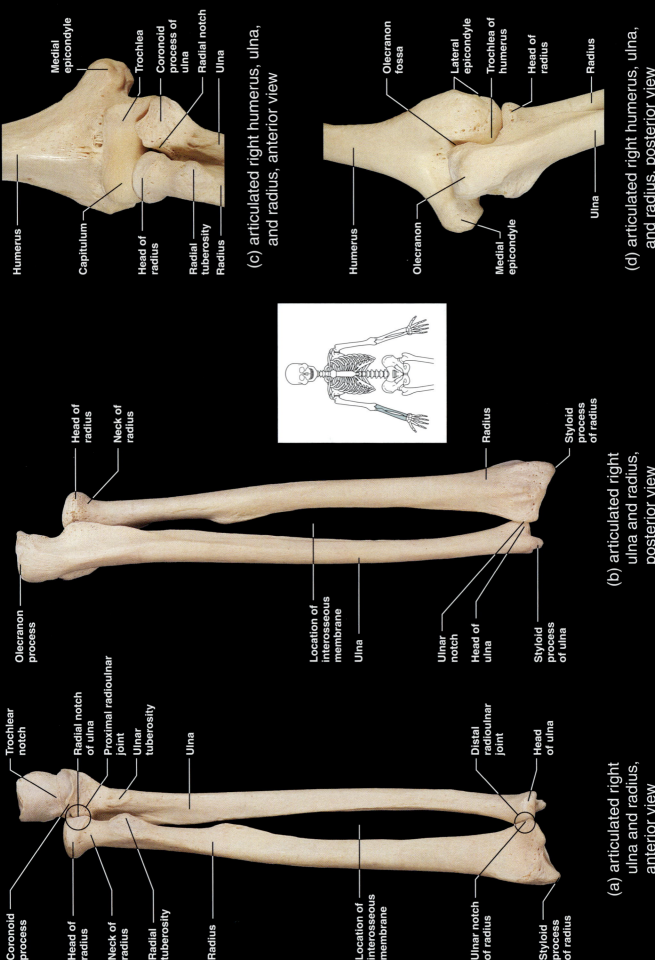

Medial epicondyle
Trochlea
Coronoid process of ulna
Radial notch
Ulna
Humerus
Capitulum
Head of radius
Radial tuberosity
Radius

(c) articulated right humerus, ulna, and radius, anterior view

Olecranon fossa
Lateral epicondyle
Trochlea of humerus
Head of radius
Radius
Humerus
Olecranon
Medial epicondyle
Ulna

(d) articulated right humerus, ulna, and radius, posterior view

Olecranon process
Head of radius
Neck of radius
Location of interosseous membrane
Ulna
Radius
Ulnar notch
Head of ulna
Styloid process of ulna
Styloid process of radius

(b) articulated right ulna and radius, posterior view

Trochlear notch
Coronoid process
Radial notch of ulna
Head of radius
Proximal radioulnar joint
Neck of radius
Ulnar tuberosity
Radial tuberosity
Ulna
Radius
Location of interosseous membrane
Distal radioulnar joint
Ulnar notch of radius
Head of ulna
Styloid process of radius

(a) articulated right ulna and radius, anterior view

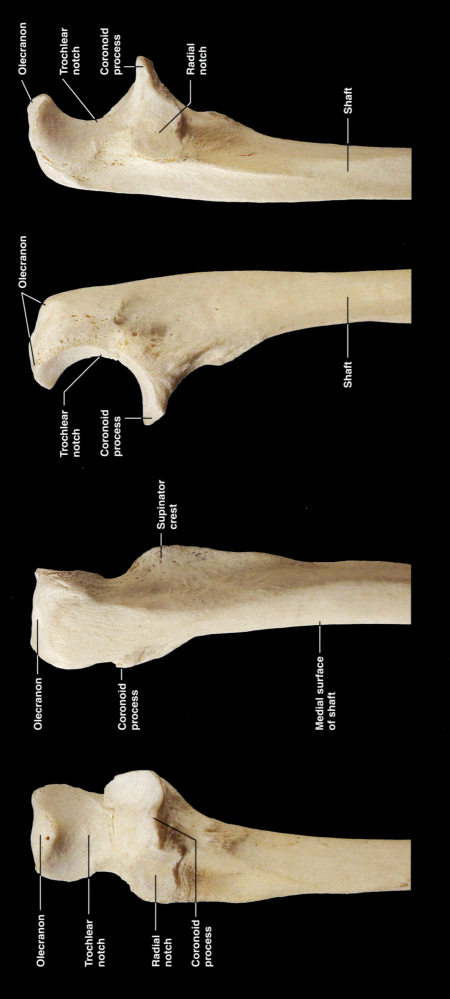

lateral view

Olecranon

Trochlear
notch

Coronoid
process

Radial
notch

Shaft

medial view

Olecranon

Trochlear
notch

Coronoid
process

Shaft

posterior view

Supinator
crest

Olecranon

Coronoid
process

Medial surface
of shaft

anterior view

Olecranon

Trochlear
notch

Radial
notch

Coronoid
process

(e) right ulna, proximal end

Figure 26 Right ulna and radius.

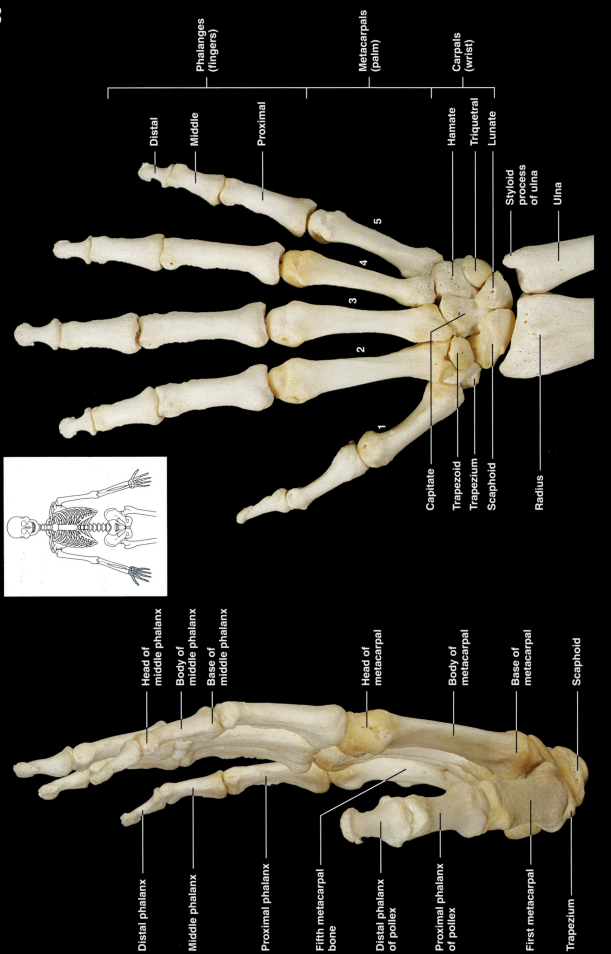

Distal phalanx

Middle phalanx

Proximal phalanx

Fifth metacarpal bone

Distal phalanx of pollex

Proximal phalanx of pollex

First metacarpal

Trapezium

Head of middle phalanx

Body of middle phalanx

Base of middle phalanx

Head of metacarpal

Body of metacarpal

Base of metacarpal

Scaphoid

(a) lateral aspect

Phalanges (fingers)

Distal

Middle

Proximal

Metacarpals (palm)

Carpals (wrist)

Hamate

Triquetral

Lunate

Styloid process of ulna

Ulna

Capitate

Trapezoid

Trapezium

Scaphoid

Radius

1

2

3

4

5

(b) dorsal aspect

Figure 27 Bones of the right hand.

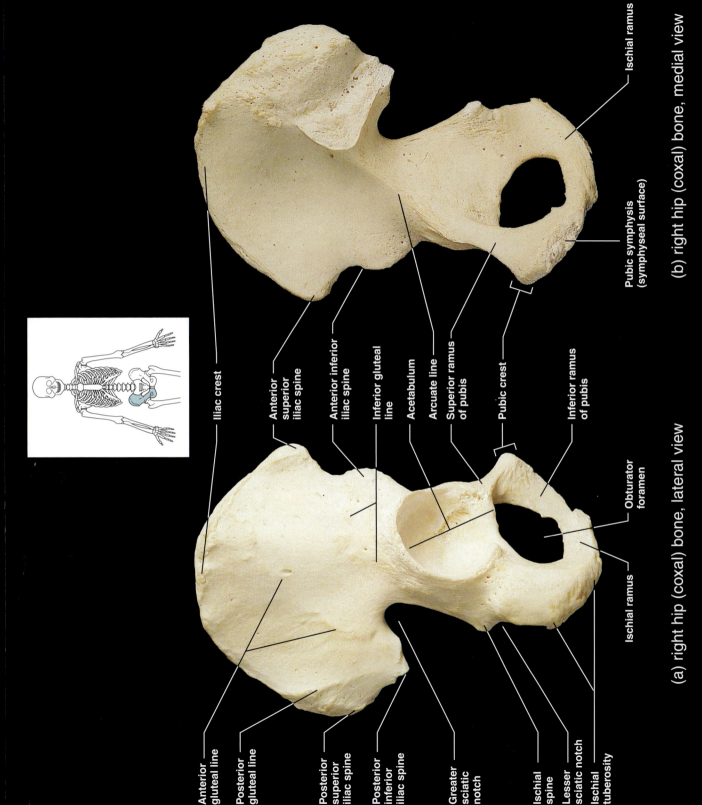

Figure 28 Bones of the male pelvis.

(a) right hip (coxal) bone, lateral view

(b) right hip (coxal) bone, medial view

Iliac crest

Anterior superior iliac spine

Anterior inferior iliac spine

Inferior gluteal line

Acetabulum

Arcuate line

Superior ramus of pubis

Pubic crest

Inferior ramus of pubis

Obturator foramen

Ischial ramus

Pubic symphysis (symphyseal surface)

Ischial ramus

Anterior gluteal line

Posterior gluteal line

Posterior superior iliac spine

Posterior inferior iliac spine

Greater sciatic notch

Ischial spine

Lesser sciatic notch

Ischial tuberosity

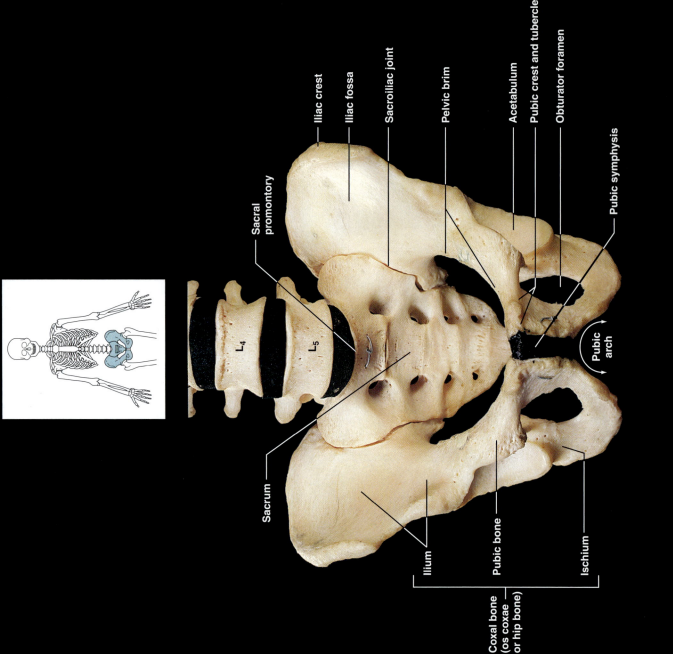

Iliac crest

Iliac fossa

Sacroiliac joint

Pelvic brim

Acetabulum

Pubic crest and tubercle

Obturator foramen

Pubic symphysis

Sacral promontory

L4

L5

Pubic arch

Sacrum

Ilium

Pubic bone

Ischium

Coxal bone (os coxae or hip bone)

(c) articulated male pelvis, anterior view

69

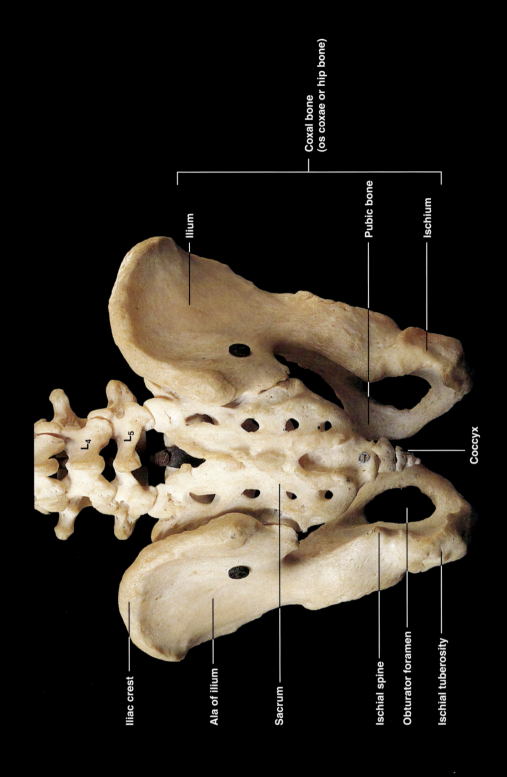

Iliac crest

Ala of ilium

Sacrum

Ischial spine

Obturator foramen

Ischial tuberosity

Ilium

Coxal bone
(os coxae or hip bone)

Pubic bone

Ischium

Coccyx

L₄

L₅

(d) articulated male pelvis, posterior view

Figure 28 Bones of the male pelvis (continued).

Greater trochan...

Intertro... crest

Gluteal tuberos...

Medial a... supraco...

Lateral e...
Lateral c...

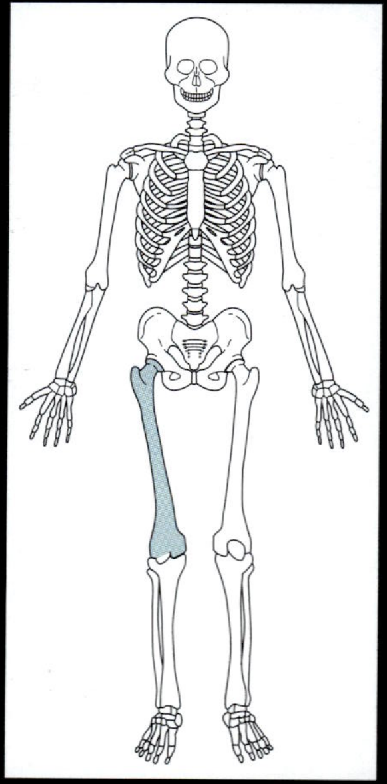

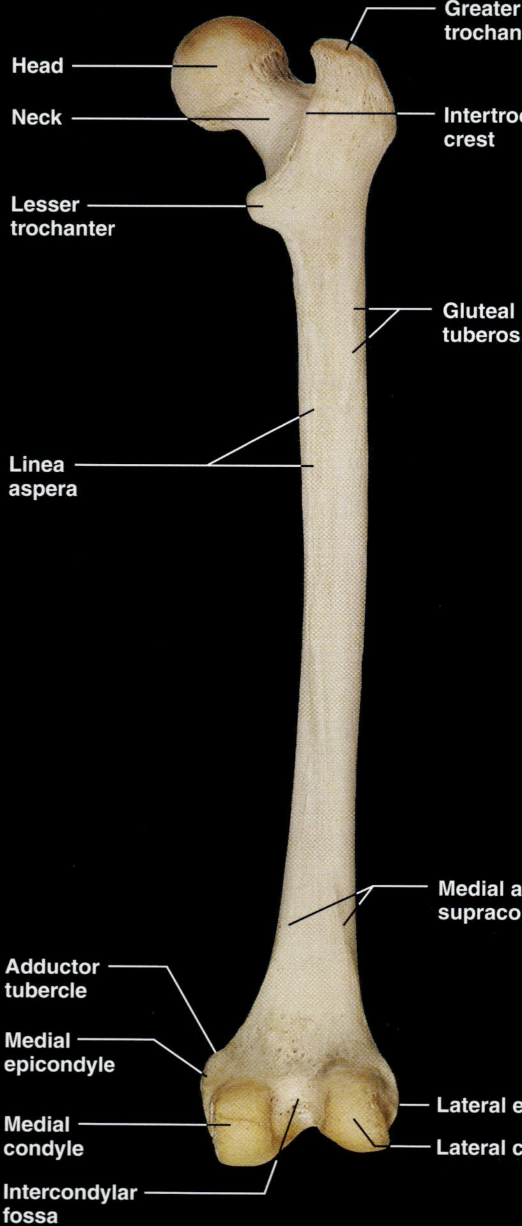

Head

Neck

Lesser trochanter

Linea aspera

Adductor tubercle

Medial epicondyle

Medial condyle

Intercondylar fossa

Head

Intertrochanteric line

Lesser trochanter

Adductor tubercle

Medial epicondyle

Medial condyle

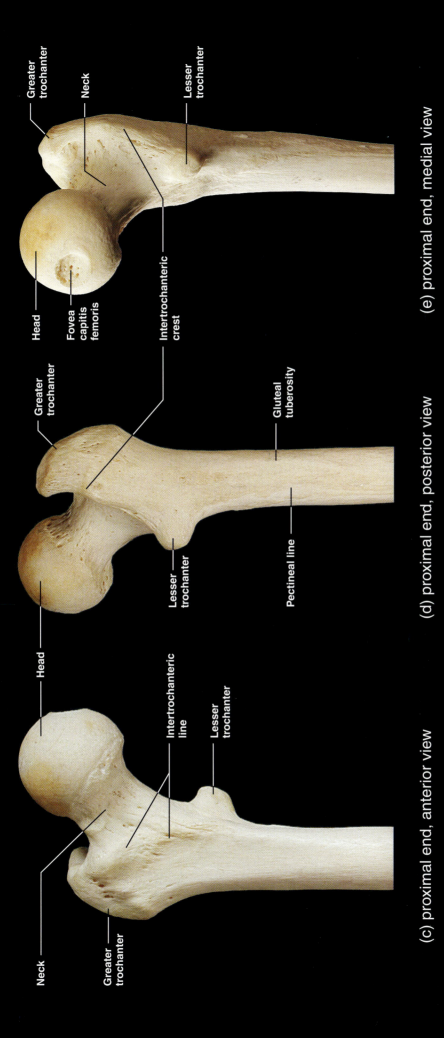

(c) proximal end, anterior view

(d) proximal end, posterior view

(e) proximal end, medial view

Figure 29 Right femur.

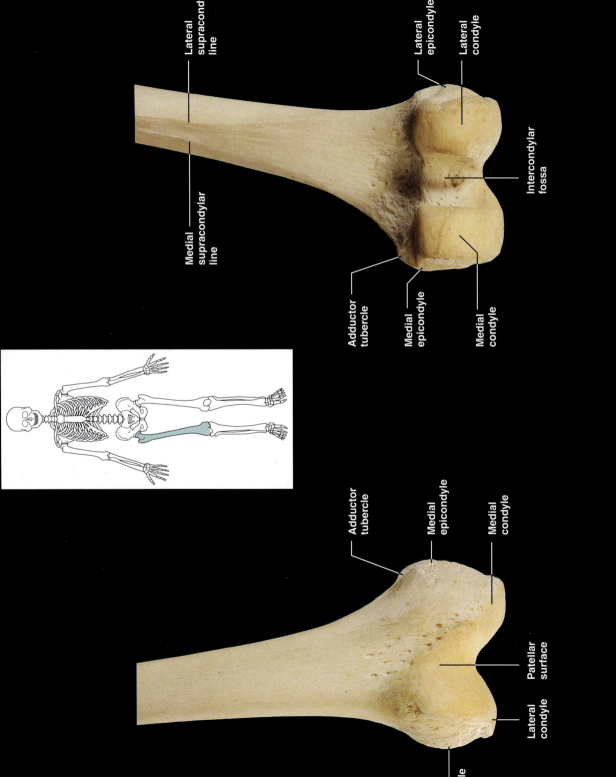

Lateral
supracondylar
line

Medial
supracondylar
line

Lateral
epicondyle

Lateral
condyle

Intercondylar
fossa

Adductor
tubercle

Medial
epicondyle

Medial
condyle

(g) distal end, posterior view

Adductor
tubercle

Medial
epicondyle

Medial
condyle

Patellar
surface

Lateral
condyle

(f) distal end, anterior view

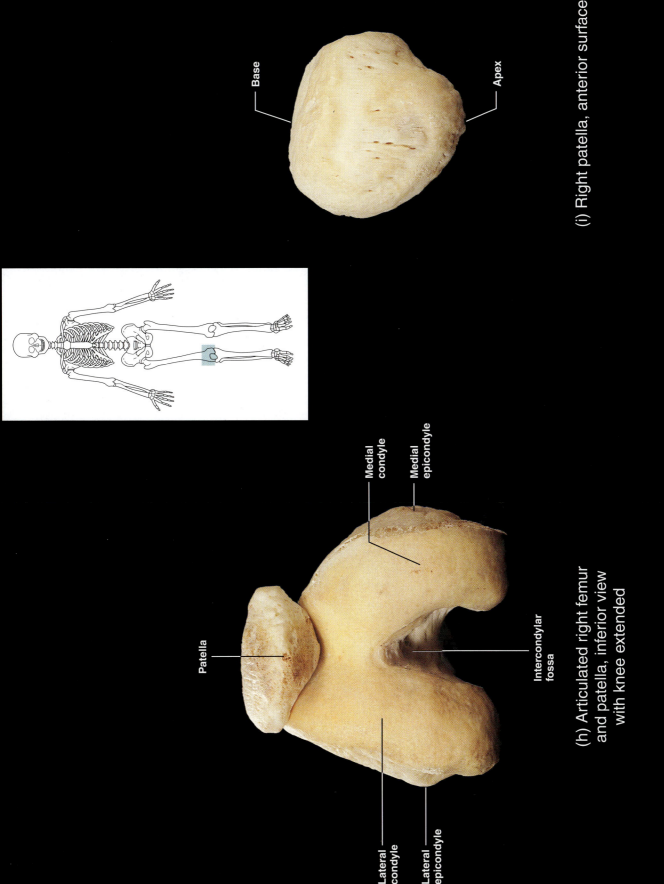

Base

Apex

(i) Right patella, anterior surface

Medial condyle

Medial epicondyle

Patella

Intercondylar fossa

Lateral condyle

Lateral epicondyle

(h) Articulated right femur and patella, inferior view with knee extended

Figure 29 Right femur (continued).

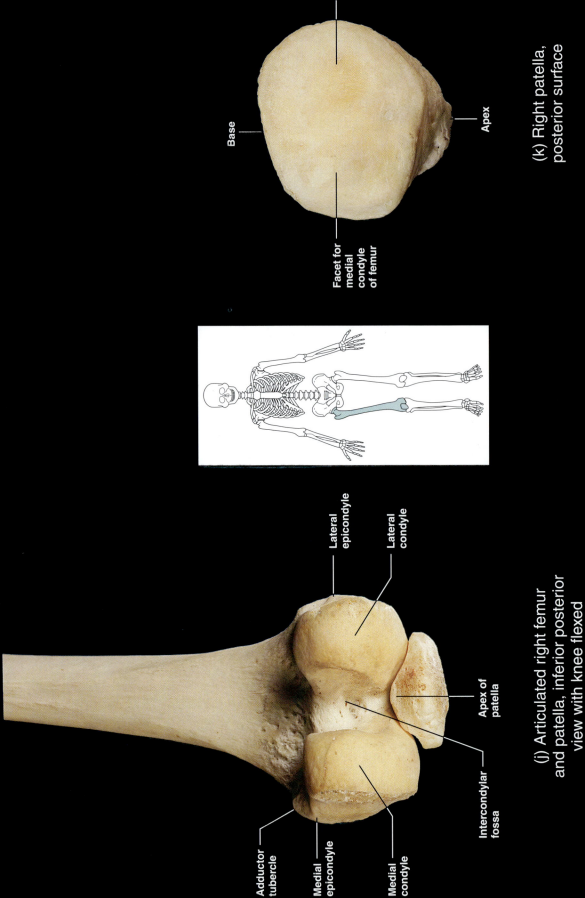

Base

Facet for medial condyle of femur

Apex

(k) Right patella, posterior surface

Lateral epicondyle

Lateral condyle

Apex of patella

Intercondylar fossa

Adductor tubercle

Medial epicondyle

Medial condyle

(j) Articulated right femur and patella, inferior posterior view with knee flexed

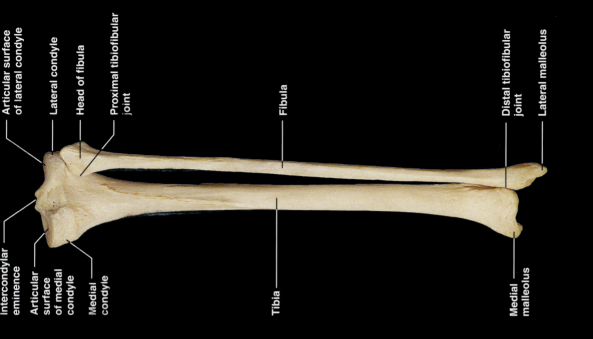

Articular surface of lateral condyle

Lateral condyle

Head of fibula

Proximal tibiofibular joint

Fibula

Distal tibiofibular joint

Lateral malleolus

Intercondylar eminence

Articular surface of medial condyle

Medial condyle

Tibia

Medial malleolus

(b) articulated right tibia and fibula, posterior view

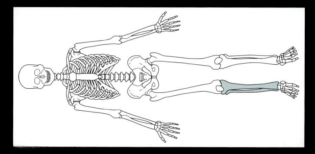

Medial condyle of tibia

Tibial tuberosity

Interosseous border of tibia

Anterior border

Tibia

Medial malleolus

Inferior articular surface

Lateral condyle of tibia

Head of fibula

Anterior border of fibula

Fibula

Lateral malleolus of fibula

(a) articulated right tibia and fibula, anterior view

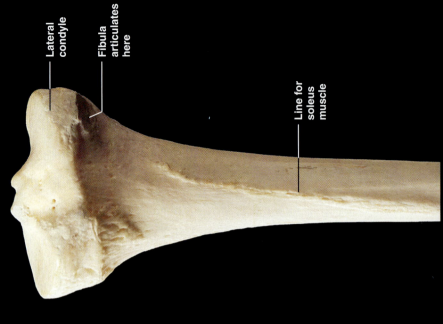

Lateral condyle

Fibula articulates here

Line for soleus muscle

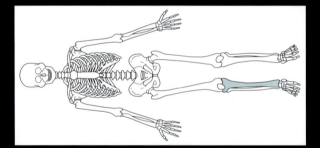

Lateral condyle

Tibial tuberosity

Fibula

Distal
tibiofibular
joint

Lateral
malleolus

Tibia

Medial
malleolus

(g) articulated right tibia
and fibula, distal end,
posterior view

Lateral
condyle

Head of
fibula

Proximal
tibiofibular
joint

Fibula

Intercondylar
eminence

Medial
condyle

Tibia

(f) articulated right tibia
and fibula, proximal end,
posterior view

Figure 30 Right tibia and fibula (continued).

Anterior
intercondylar
area

Tibial
tuberosity

Lateral
condyle

Posterior
intercondylar
area

Intercondylar
eminence

Medial
condyle

(e) right tibia, proximal end,
articular surface

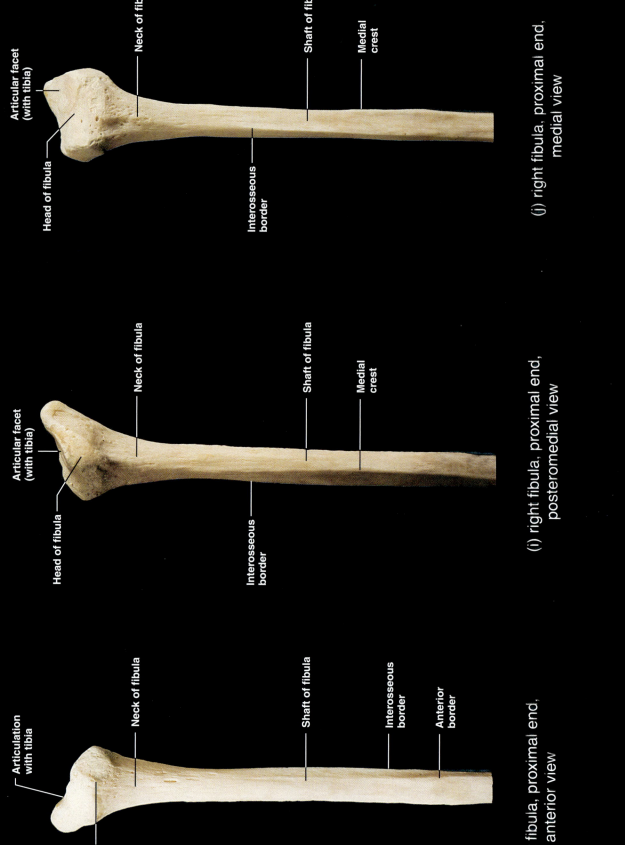

Articular facet
(with tibia)

Neck of fib

Head of fibula

Shaft of fib

Medial
crest

Interosseous
border

(j) right fibula, proximal end,
medial view

Articular facet
(with tibia)

Neck of fibula

Head of fibula

Shaft of fibula

Medial
crest

Interosseous
border

(i) right fibula, proximal end,
posteromedial view

Articulation
with tibia

Neck of fibula

Shaft of fibula

Interosseous
border

Anterior
border

fibula, proximal end,
anterior view

Figure 30 Right tibia and fibula (continued).

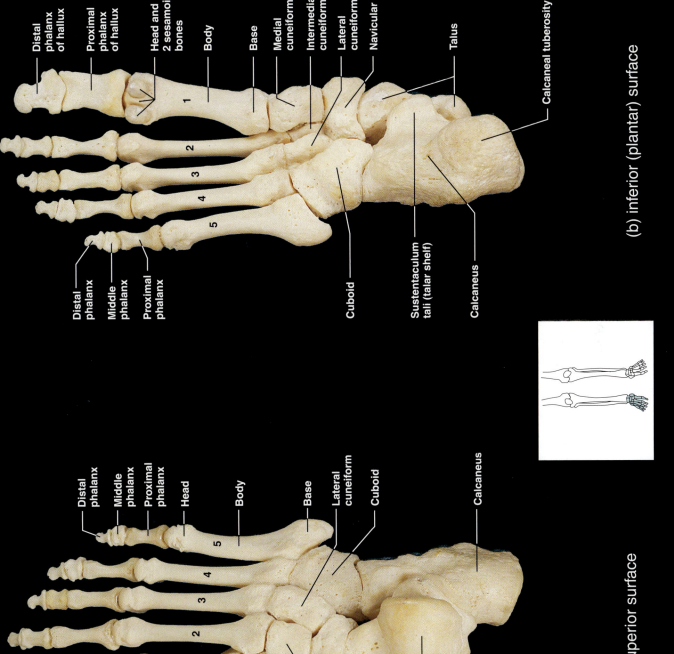

Figure 31 Bones of the right ankle and foot.

Distal phalanx of hallux

Proximal phalanx of hallux

Head and 2 sesamoid bones

Body

Base

Medial cuneiform

Intermediate cuneiform

Lateral cuneiform

Navicular

Talus

Calcaneal tuberosity

Distal phalanx

Middle phalanx

Proximal phalanx

Cuboid

Sustentaculum tali (talar shelf)

Calcaneus

(b) inferior (plantar) surface

Distal phalanx

Middle phalanx

Proximal phalanx

Head

Body

Base

Lateral cuneiform

Cuboid

Calcaneus

superior surface

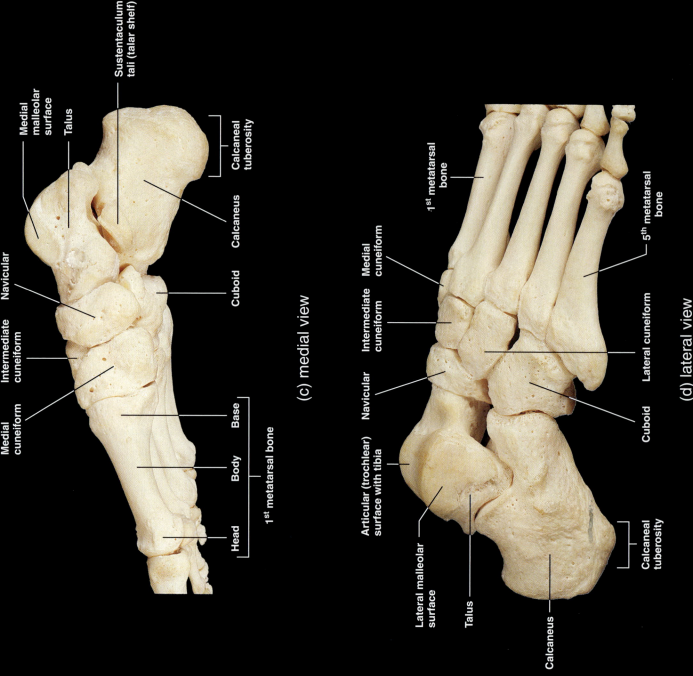

Medial malleolar surface

Talus

Sustentaculum tali (talar shelf)

Calcaneal tuberosity

Navicular

Calcaneus

Intermediate cuneiform

Cuboid

Medial cuneiform

Base

Body

Head

1st metatarsal bone

(c) medial view

1st metatarsal bone

Medial cuneiform

Intermediate cuneiform

5th metatarsal bone

Navicular

Lateral cuneiform

Articular (trochlear) surface with tibia

Cuboid

Lateral malleolar surface

Talus

Calcaneal tuberosity

Calcaneus

(d) lateral view

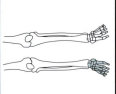

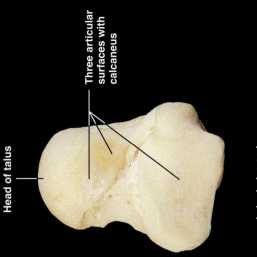

Three articular
surfaces with
calcaneus

Head of talus

(g) right talus,
inferior view

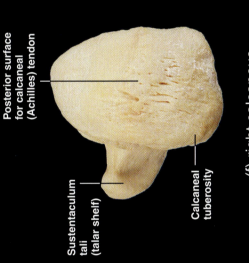

Posterior surface
for calcaneal
(Achilles) tendon

Sustentaculum
tali (talar shelf)

Calcaneal
tuberosity

(f) right calcaneus,
posterior aspect

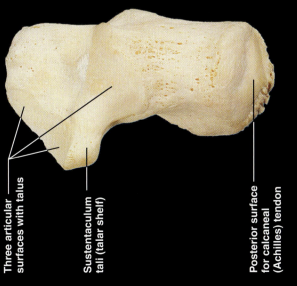

Three articular
surfaces with talus

Sustentaculum
tali (talar shelf)

Posterior surface
for calcaneal
(Achilles) tendon

(e) right calcaneus,
superior aspect

Figure 31 Bones of the right ankle and foot (continued).

Part III

SOFT TISSUE OF THE HUMAN BODY

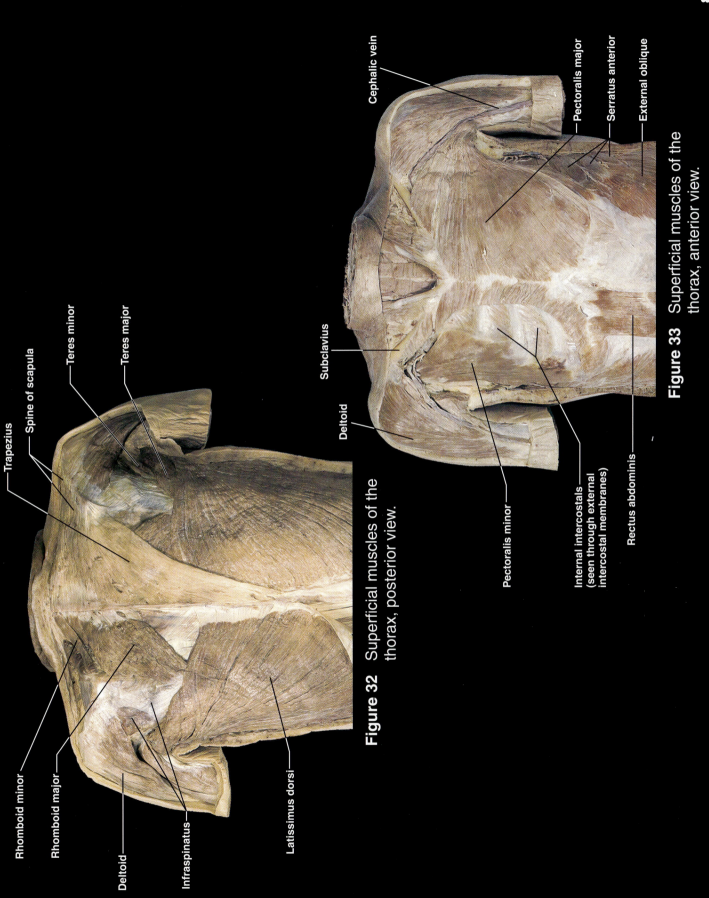

Figure 32 Superficial muscles of the thorax, posterior view.

Trapezius

Spine of scapula

Teres minor

Teres major

Rhomboid minor

Rhomboid major

Deltoid

Infraspinatus

Latissimus dorsi

Figure 33 Superficial muscles of the thorax, anterior view.

Cephalic vein

Pectoralis major

Serratus anterior

External oblique

Subclavius

Deltoid

Pectoralis minor

Internal intercostals (seen through external intercostal membranes)

Rectus abdominis

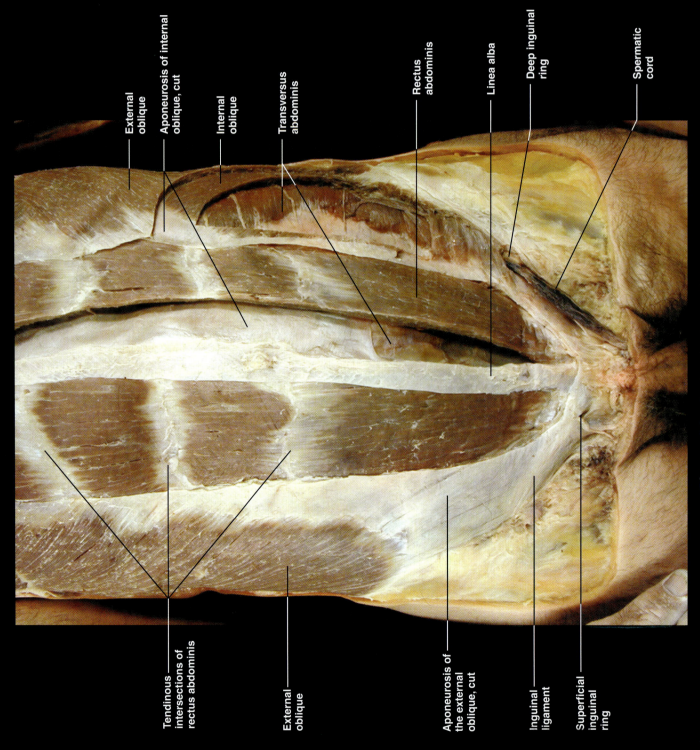

External oblique

Aponeurosis of internal oblique, cut

Internal oblique

Transversus abdominis

Rectus abdominis

Linea alba

Deep inguinal ring

Spermatic cord

Tendinous intersections of rectus abdominis

External oblique

Aponeurosis of the external oblique, cut

Inguinal ligament

Superficial inguinal ring

Figure 34 Abdominal muscles.

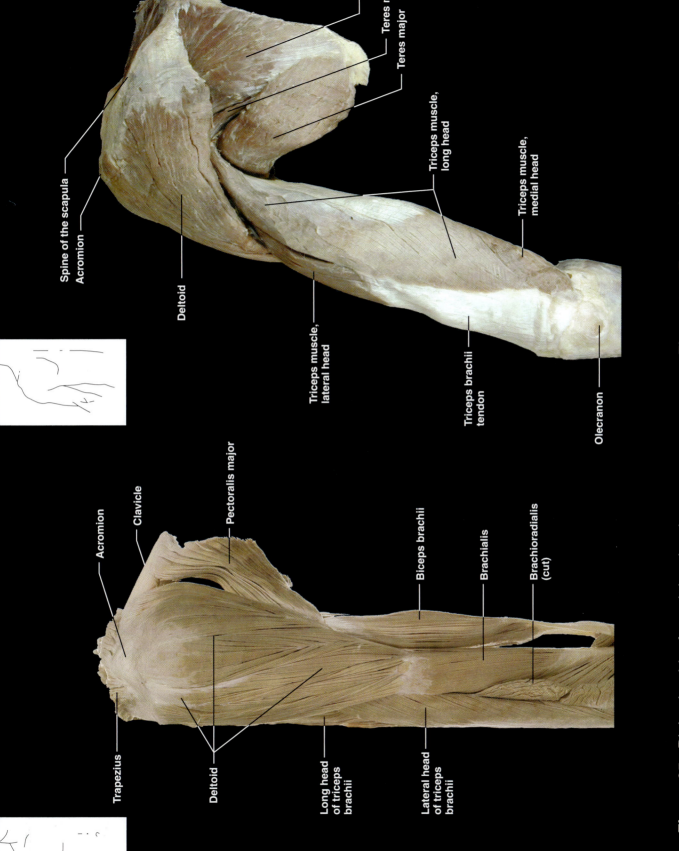

Spine of the scapula
Acromion

Deltoid

Triceps muscle, lateral head

Triceps brachii tendon

Olecranon

Teres
Teres major

Triceps muscle, long head

Triceps muscle, medial head

Triceps muscle, long head

Triceps muscle, medial head

Figure 36 Triceps of the left arm, posterior view.

Acromion

Clavicle

Pectoralis major

Biceps brachii

Brachialis

Brachioradialis (cut)

Trapezius

Deltoid

Long head of triceps brachii

Lateral head of triceps brachii

Figure 35 Right shoulder from right, showing deltoid muscle and biceps.

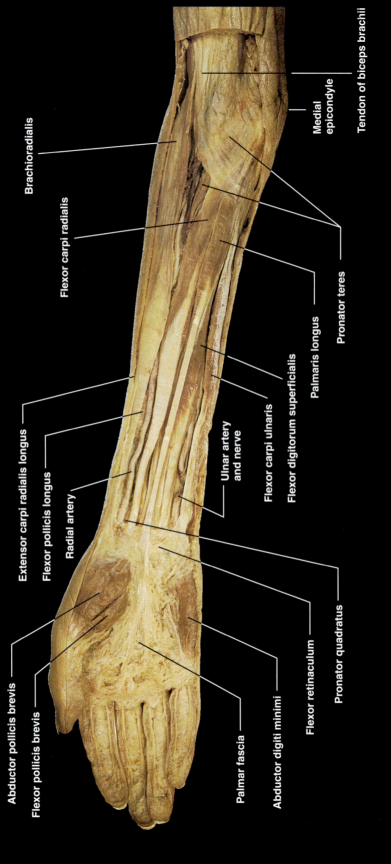

Brachioradialis

Flexor carpi radialis

Extensor carpi radialis longus

Flexor pollicis longus

Radial artery

Abductor pollicis brevis

Flexor pollicis brevis

Palmar fascia

Abductor digiti minimi

Flexor retinaculum

Pronator quadratus

Ulnar artery
and nerve

Flexor carpi ulnaris

Flexor digitorum superficialis

Palmaris longus

Pronator teres

Medial
epicondyle

Tendon of biceps brachii

(a) palmar surface

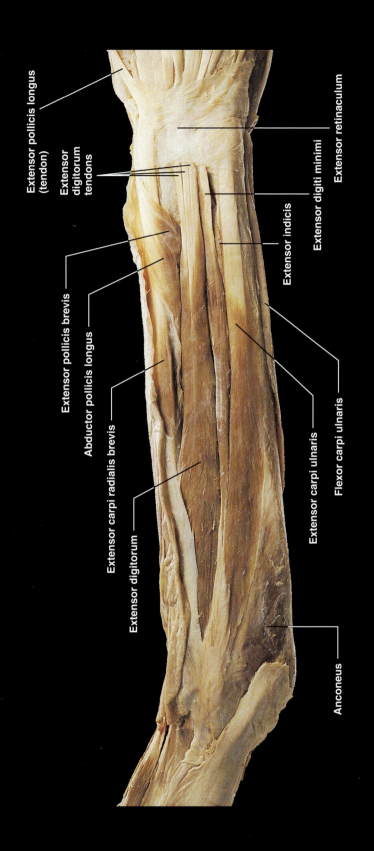

Extensor pollicis longus (tendon)

Extensor digitorum tendons

Extensor retinaculum

Extensor indicis

Extensor digiti minimi

Extensor pollicis brevis

Abductor pollicis longus

Extensor carpi radialis brevis

Extensor carpi ulnaris

Flexor carpi ulnaris

Extensor digitorum

Anconeus

(b) dorsum surface

Figure 37 Right forearm and wrist.

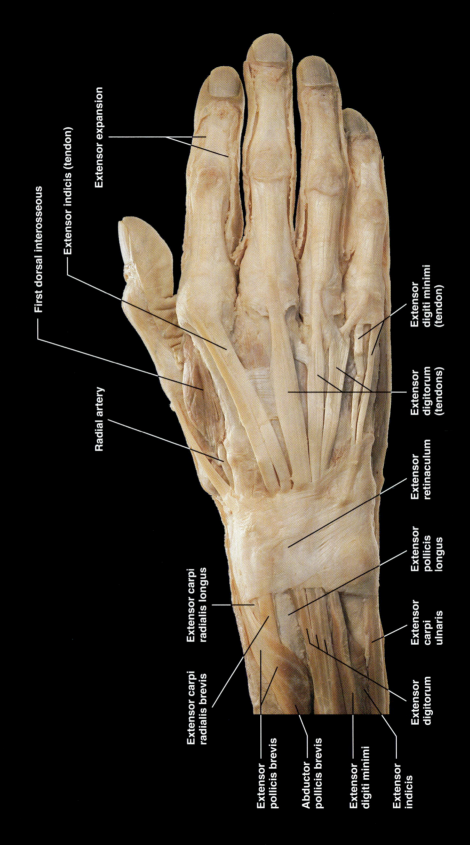

First dorsal interosseous

Extensor indicis (tendon)

Extensor expansion

Radial artery

Extensor digiti minimi (tendon)

Extensor digitorum (tendons)

Extensor retinaculum

Extensor pollicis longus

Extensor carpi ulnaris

Extensor digitorum

Extensor indicis

Extensor digiti minimi

Abductor pollicis brevis

Extensor pollicis brevis

Extensor carpi radialis longus

Extensor carpi radialis brevis

(a) dorsum surface of the right hand and wrist

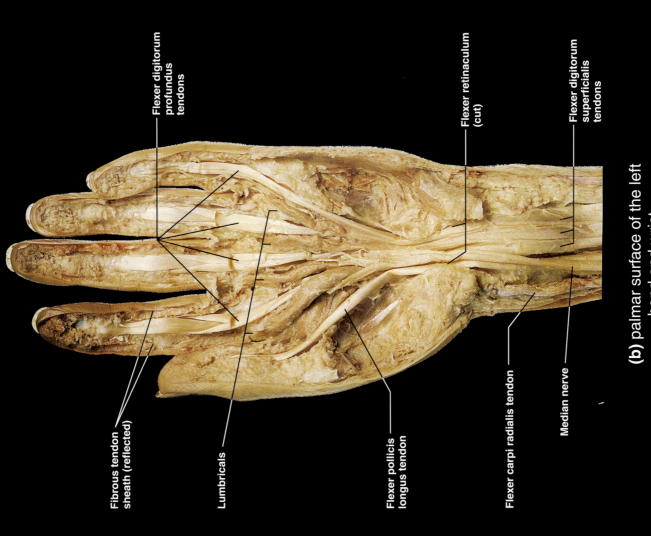

Flexer digitorum profundus tendons

Flexer retinaculum (cut)

Flexer digitorum superficialis tendons

Fibrous tendon sheath (reflected)

Lumbricals

Flexer pollicis longus tendon

Flexer carpi radialis tendon

Median nerve

(b) palmar surface of the left hand and wrist

Figure 38 Wrist and hand.

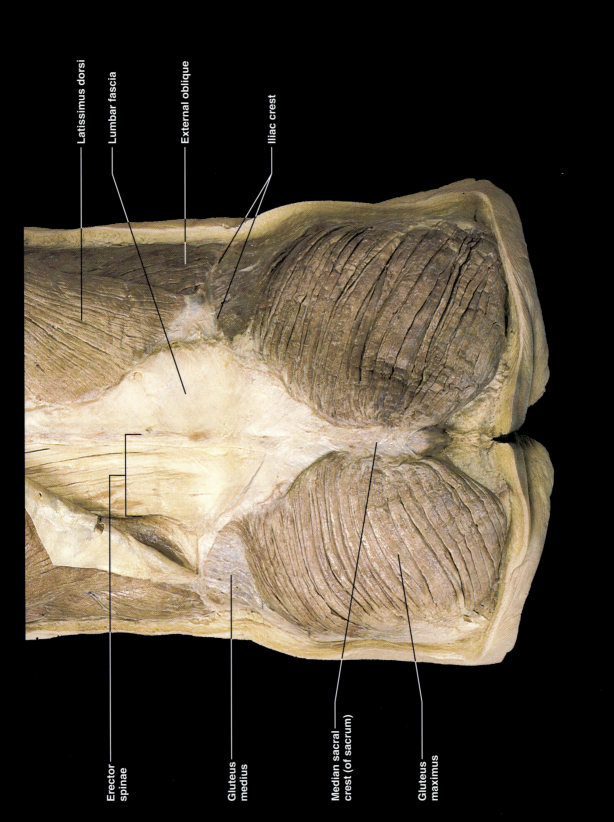

Latissimus dorsi

Lumbar fascia

External oblique

Iliac crest

Erector
spinae

Gluteus
medius

Median sacral
crest (of sacrum)

Gluteus
maximus

Figure 39 Superficial muscles of the superior gluteal region.

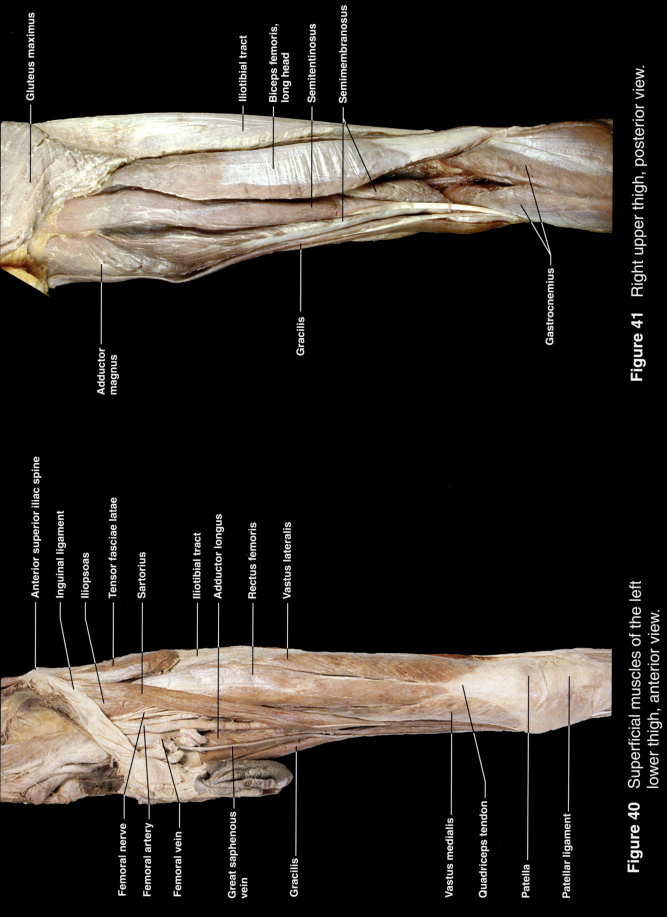

Figure 41 Right upper thigh, posterior view.

Gluteus maximus

Iliotibial tract

Biceps femoris, long head

Semitentinosus

Semimembranosus

Adductor magnus

Gracilis

Gastrocnemius

Figure 40 Superficial muscles of the left lower thigh, anterior view.

Anterior superior iliac spine

Inguinal ligament

Iliopsoas

Tensor fasciae latae

Sartorius

Iliotibial tract

Adductor longus

Rectus femoris

Vastus lateralis

Femoral nerve

Femoral artery

Femoral vein

Great saphenous vein

Gracilis

Vastus medialis

Quadriceps tendon

Patella

Patellar ligament

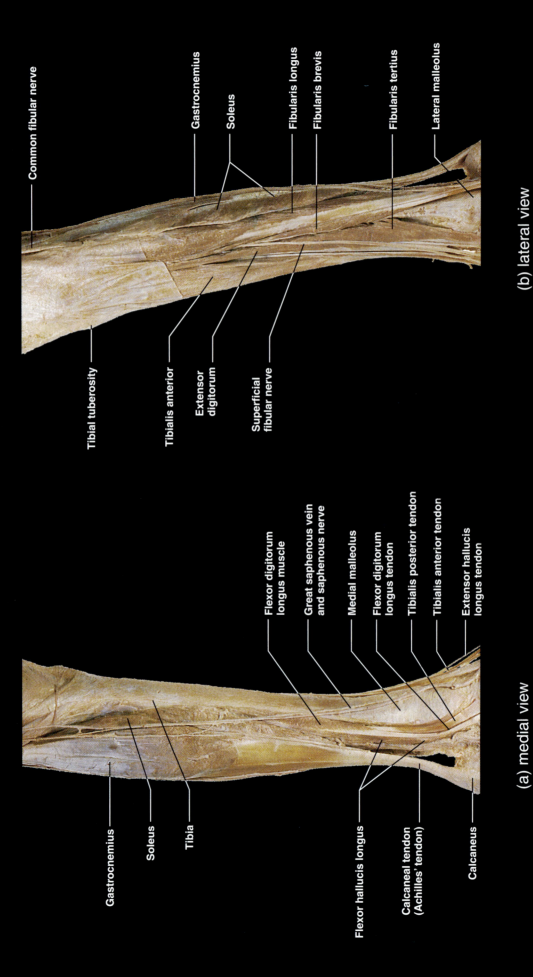

Common fibular nerve

Gastrocnemius

Soleus

Fibularis longus

Fibularis brevis

Fibularis tertius

Lateral malleolus

Tibial tuberosity

Tibialis anterior

Extensor digitorum

Superficial fibular nerve

(b) lateral view

Flexor digitorum longus muscle

Great saphenous vein and saphenous nerve

Medial malleolus

Flexor digitorum longus tendon

Tibialis posterior tendon

Tibialis anterior tendon

Extensor hallucis longus tendon

Gastrocnemius

Soleus

Tibia

Flexor hallucis longus

Calcaneal tendon (Achilles' tendon)

Calcaneus

(a) medial view

Figure 42 Leg.

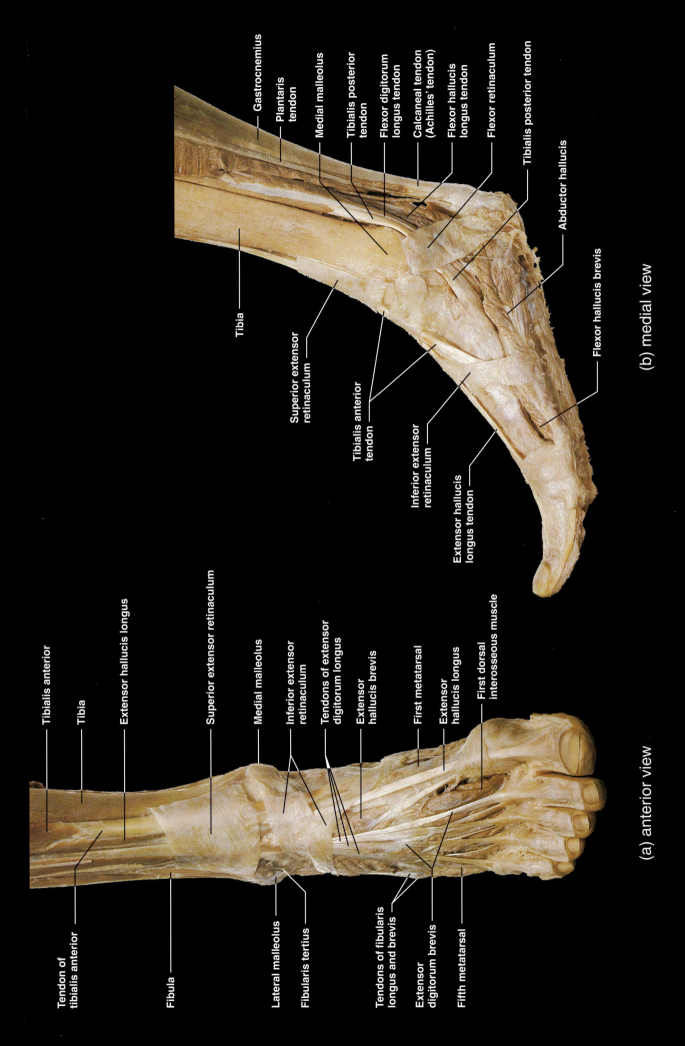

Figure 43 Foot.

(a) anterior view

(b) medial view

Tendon of tibialis anterior

Fibula

Lateral malleolus

Fibularis tertius

Tendons of fibularis longus and brevis

Extensor digitorum brevis

Fifth metatarsal

Tibialis anterior

Tibia

Extensor hallucis longus

Superior extensor retinaculum

Medial malleolus

Inferior extensor retinaculum

Tendons of extensor digitorum longus

Extensor hallucis brevis

First metatarsal

Extensor hallucis longus

First dorsal interosseous muscle

Tibia

Superior extensor retinaculum

Tibialis anterior tendon

Inferior extensor retinaculum

Extensor hallucis longus tendon

Flexor hallucis brevis

Gastrocnemius

Plantaris tendon

Medial malleolus

Tibialis posterior tendon

Flexor digitorum longus tendon

Calcaneal tendon (Achilles' tendon)

Flexor hallucis longus tendon

Flexor retinaculum

Tibialis posterior tendon

Abductor hallucis

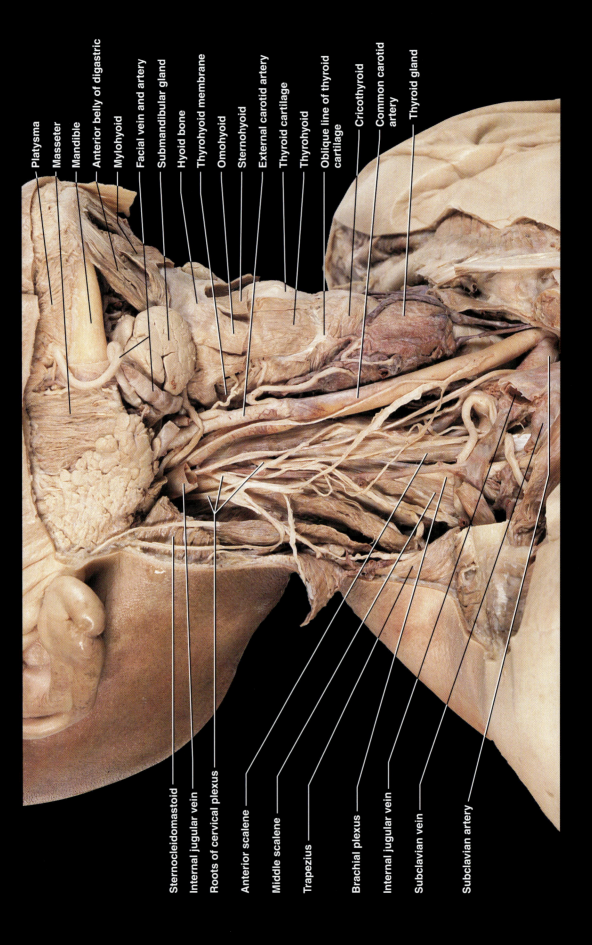

Platysma
Masseter
Mandible
Anterior belly of digastric
Mylohyoid
Facial vein and artery
Submandibular gland
Hyoid bone
Thyrohyoid membrane
Omohyoid
Sternohyoid
External carotid artery
Thyroid cartilage
Thyrohyoid
Oblique line of thyroid cartilage
Cricothyroid
Common carotid artery
Thyroid gland

Sternocleidomastoid
Internal jugular vein
Roots of cervical plexus
Anterior scalene
Middle scalene
Trapezius
Brachial plexus
Internal jugular vein
Subclavian vein
Subclavian artery

Figure 44 Right lower face and upper neck.

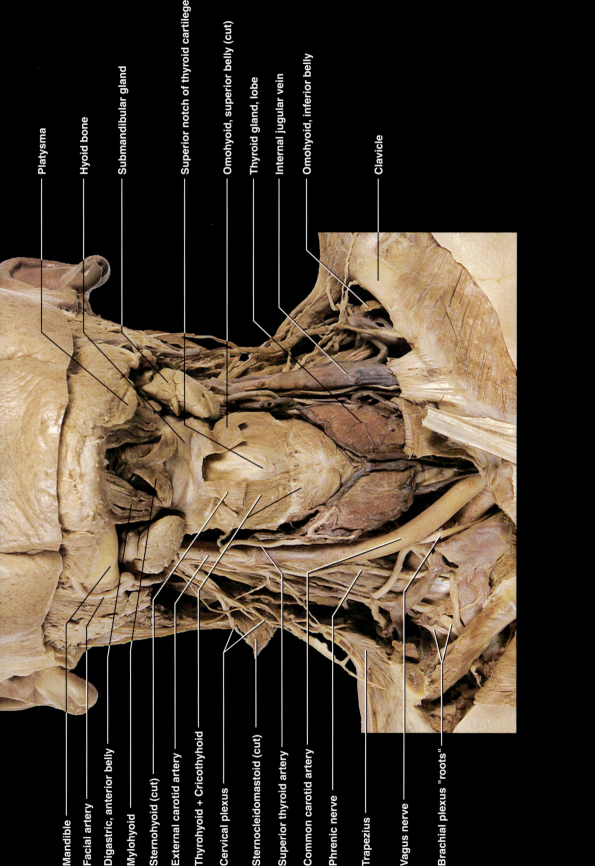

Platysma

Hyoid bone

Submandibular gland

Superior notch of thyroid cartilege

Omohyoid, superior belly (cut)

Thyroid gland, lobe

Internal jugular vein

Omohyoid, inferior belly

Clavicle

Mandible

Facial artery

Digastric, anterior belly

Mylohyoid

Sternohyoid (cut)

External carotid artery

Thyrohyoid + Cricothyhoid

Cervical plexus

Sternocleidomastoid (cut)

Superior thyroid artery

Common carotid artery

Phrenic nerve

Trapezius

Vagus nerve

Brachial plexus "roots"

Figure 45 Muscles, blood vessels, and nerves of neck, anterior view.

97

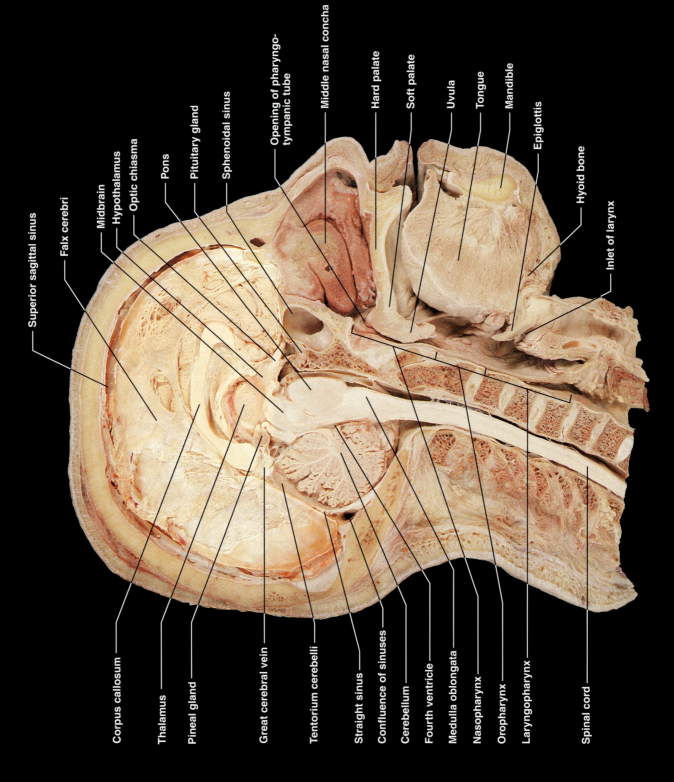

Superior sagittal sinus

Falx cerebri

Midbrain

Hypothalamus

Optic chiasma

Pons

Pituitary gland

Sphenoidal sinus

Opening of pharyngo-tympanic tube

Middle nasal concha

Hard palate

Soft palate

Uvula

Tongue

Mandible

Epiglottis

Hyoid bone

Inlet of larynx

Corpus callosum

Thalamus

Pineal gland

Great cerebral vein

Tentorium cerebelli

Straight sinus

Confluence of sinuses

Cerebellum

Fourth ventricle

Medulla oblongata

Nasopharynx

Oropharynx

Laryngopharynx

Spinal cord

Figure 46 Sagittal section of the head.

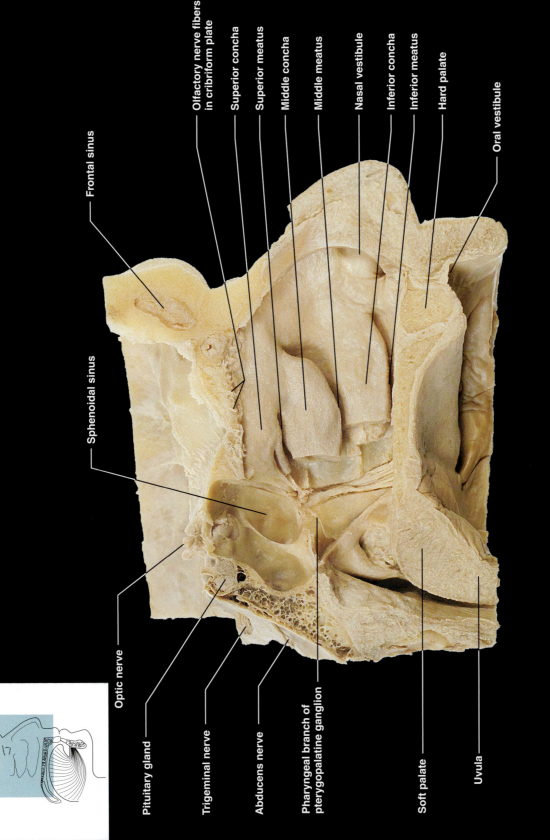

Frontal sinus

Olfactory nerve fibers in cribriform plate

Superior concha

Superior meatus

Middle concha

Middle meatus

Nasal vestibule

Inferior concha

Inferior meatus

Hard palate

Oral vestibule

Sphenoidal sinus

Optic nerve

Pituitary gland

Trigeminal nerve

Abducens nerve

Pharyngeal branch of pterygopalatine ganglion

Soft palate

Uvula

Figure 47 Left nasal cavity, lateral wall.

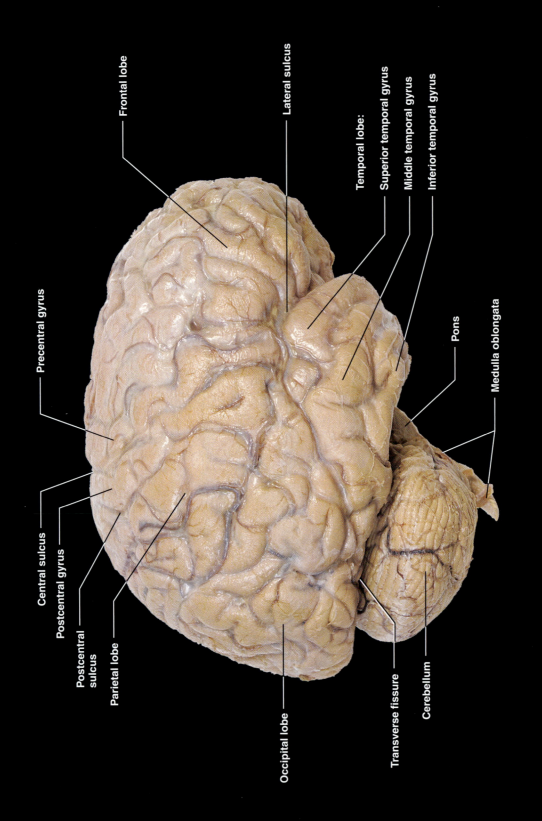

Precentral gyrus

Frontal lobe

Lateral sulcus

Temporal lobe:

Superior temporal gyrus

Middle temporal gyrus

Inferior temporal gyrus

Pons

Medulla oblongata

Central sulcus

Postcentral gyrus

Postcentral sulcus

Parietal lobe

Occipital lobe

Transverse fissure

Cerebellum

Figure 48 Right cerebral hemisphere (arachnoid mater removed).

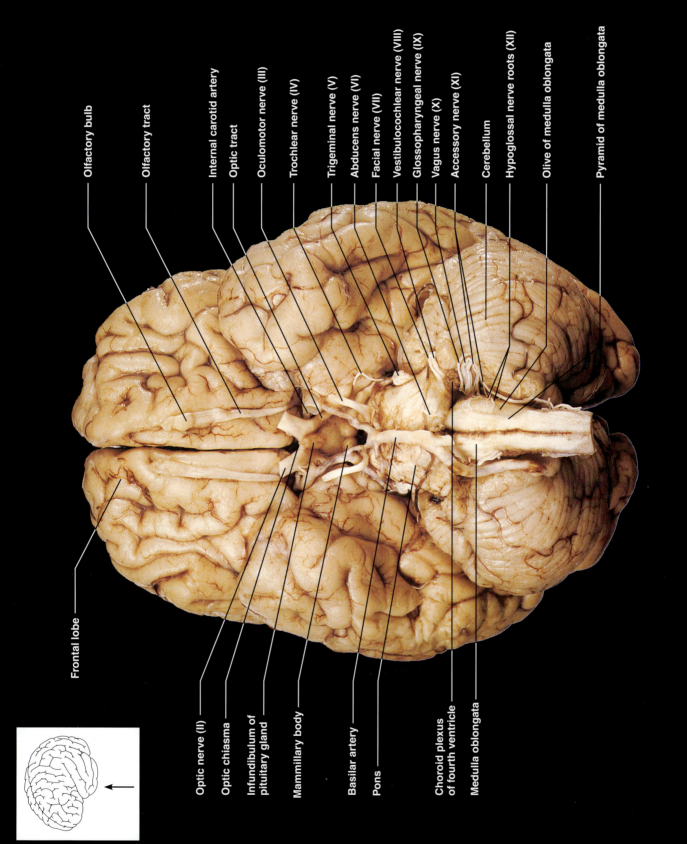

Olfactory bulb

Olfactory tract

Internal carotid artery

Optic tract

Oculomotor nerve (III)

Trochlear nerve (IV)

Trigeminal nerve (V)

Abducens nerve (VI)

Facial nerve (VII)

Vestibulocochlear nerve (VIII)

Glossopharyngeal nerve (IX)

Vagus nerve (X)

Accessory nerve (XI)

Cerebellum

Hypoglossal nerve roots (XII)

Olive of medulla oblongata

Pyramid of medulla oblongata

Frontal lobe

Optic nerve (II)

Optic chiasma

Infundibulum of pituitary gland

Mammillary body

Basilar artery

Pons

Choroid plexus of fourth ventricle

Medulla oblongata

Figure 49 Ventral view of the brain.

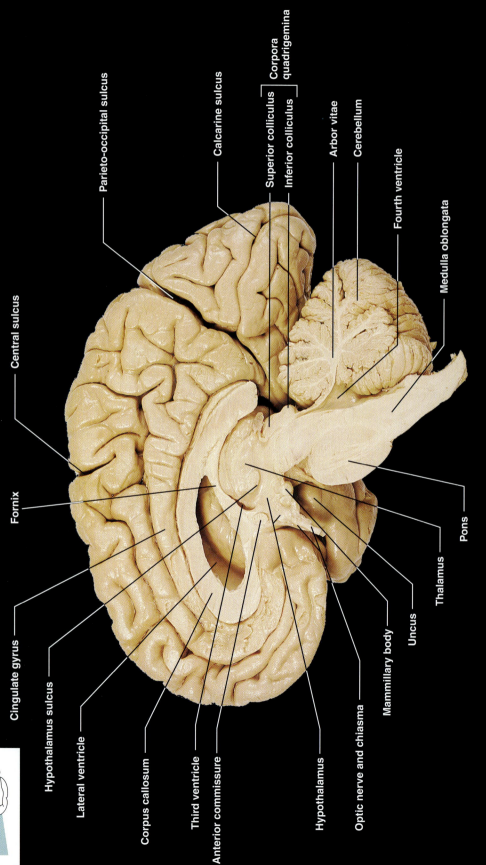

Cingulate gyrus

Hypothalamus sulcus

Lateral ventricle

Corpus callosum

Third ventricle

Anterior commissure

Hypothalamus

Optic nerve and chiasma

Mammillary body

Uncus

Thalamus

Pons

Fornix

Central sulcus

Parieto-occipital sulcus

Calcarine sulcus

Superior colliculus

Inferior colliculus

Corpora quadrigemina

Arbor vitae

Cerebellum

Fourth ventricle

Medulla oblongata

Figure 50 Midsagittal section of the brain.

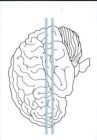

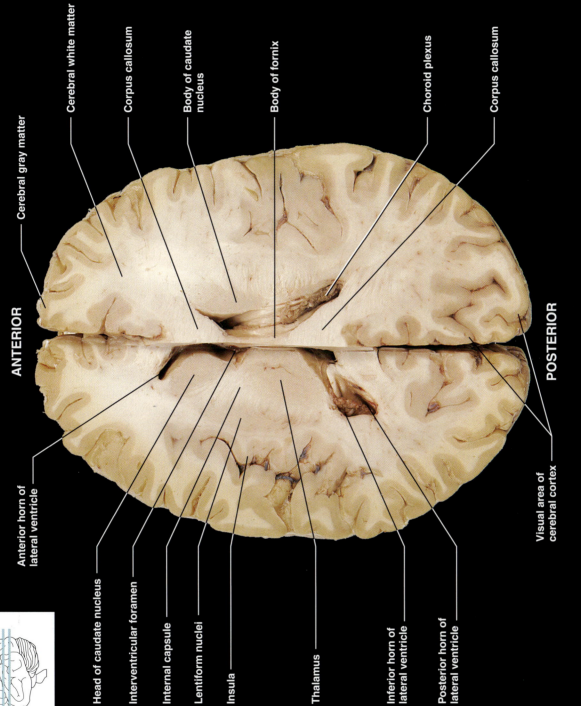

ANTERIOR

POSTERIOR

Cerebral gray matter

Cerebral white matter

Corpus callosum

Body of caudate nucleus

Body of fornix

Choroid plexus

Corpus callosum

Anterior horn of lateral ventricle

Head of caudate nucleus

Interventricular foramen

Internal capsule

Lentiform nuclei

Insula

Thalamus

Inferior horn of lateral ventricle

Posterior horn of lateral ventricle

Visual area of cerebral cortex

Figure 51 Transverse section of the brain, superior view. Left: on a level with the intraventricular foramen; right: about 1.5 cm higher.

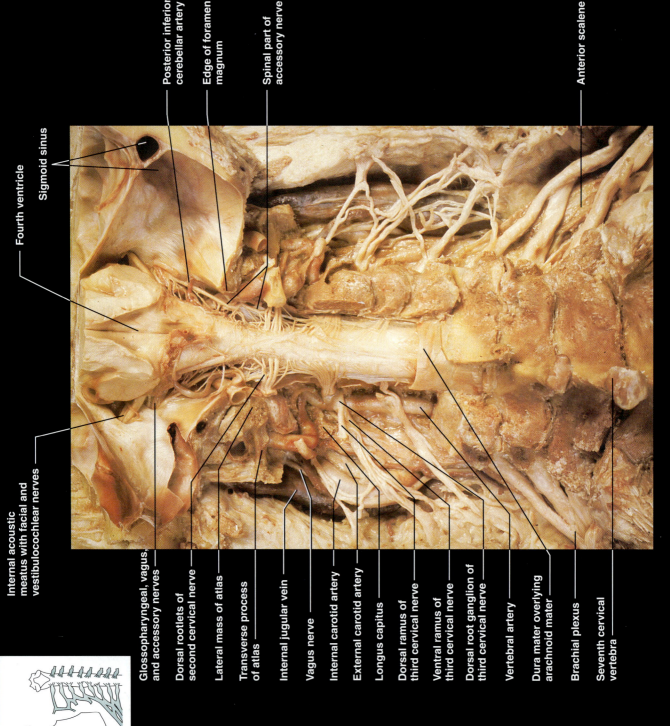

Internal acoustic meatus with facial and vestibulocochlear nerves

Fourth ventricle

Sigmoid sinus

Posterior inferior cerebellar artery

Edge of foramen magnum

Spinal part of accessory nerve

Anterior scalene

Glossopharyngeal, vagus, and accessory nerves

Dorsal rootlets of second cervical nerve

Lateral mass of atlas

Transverse process of atlas

Internal jugular vein

Vagus nerve

Internal carotid artery

External carotid artery

Longus capitus

Dorsal ramus of third cervical nerve

Ventral ramus of third cervical nerve

Dorsal root ganglion of third cervical nerve

Vertebral artery

Dura mater overlying arachnoid mater

Brachial plexus

Seventh cervical vertebra

Figure 52   Brainstem and cervical region of the spinal cord, posterior view.

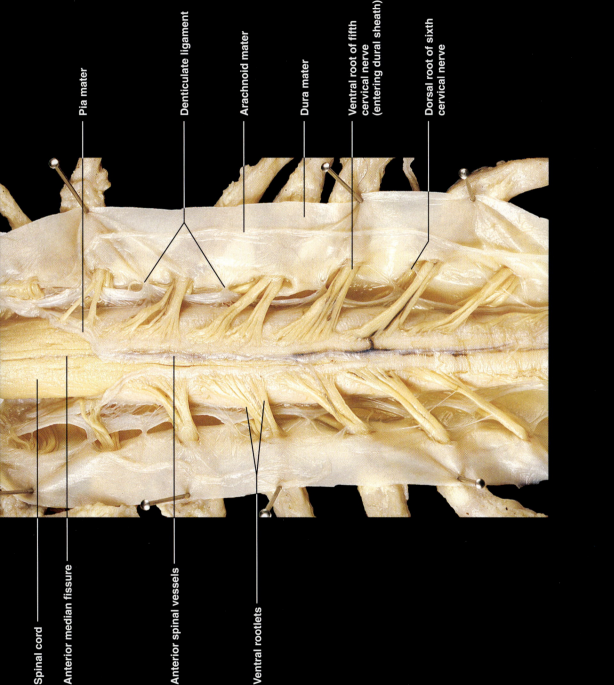

Pia mater

Denticulate ligament

Arachnoid mater

Dura mater

Ventral root of fifth
cervical nerve
(entering dural sheath)

Dorsal root of sixth
cervical nerve

Spinal cord

Anterior median fissure

Anterior spinal vessels

Ventral rootlets

Figure 53 Cervical region of spinal cord, ventral view.

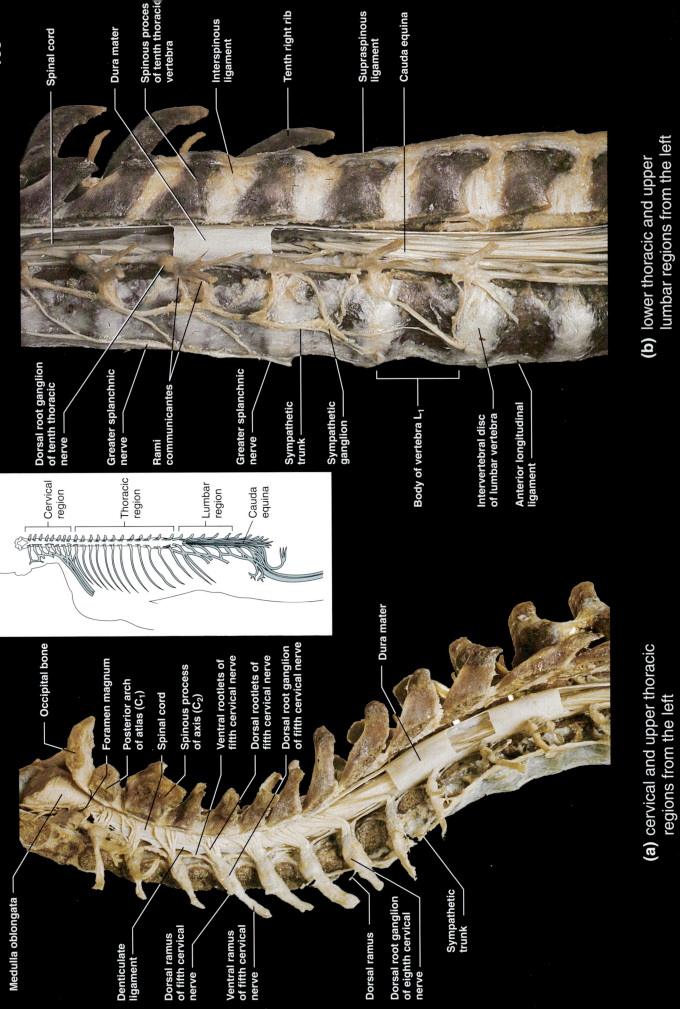

Spinal cord

Dura mater

Spinous process
of tenth thoracic
vertebra

Interspinous
ligament

Tenth right rib

Supraspinous
ligament

Cauda equina

Dorsal root ganglion
of tenth thoracic
nerve

Greater splanchnic
nerve

Rami
communicantes

Greater splanchnic
nerve

Sympathetic
trunk

Sympathetic
ganglion

Body of vertebra L₁

Intervertebral disc
of lumbar vertebra

Anterior longitudinal
ligament

(b) lower thoracic and upper
lumbar regions from the left

Cervical
region

Thoracic
region

Lumbar
region

Cauda
equina

Occipital bone

Foramen magnum

Posterior arch
of atlas (C₁)

Spinal cord

Spinous process
of axis (C₂)

Ventral rootlets of
fifth cervical nerve

Dorsal rootlets of
fifth cervical nerve

Dorsal root ganglion
of fifth cervical nerve

Dura mater

Medulla oblongata

Denticulate
ligament

Dorsal ramus
of fifth cervical
nerve

Ventral ramus
of fifth cervical
nerve

Dorsal ramus

Dorsal root ganglion
of eighth cervical
nerve

Sympathetic
trunk

(a) cervical and upper thoracic
regions from the left

Figure 54 Vertebral column and spinal cord

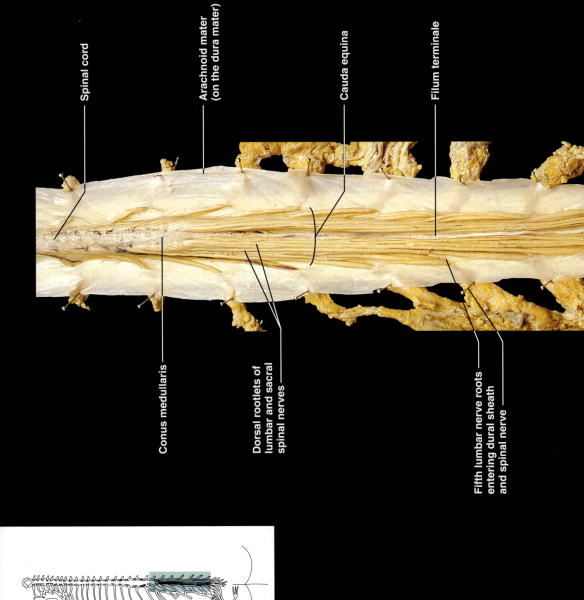

Spinal cord

Arachnoid mater
(on the dura mater)

Cauda equina

Filum terminale

Conus medullaris

Dorsal rootlets of
lumbar and sacral
spinal nerves

Fifth lumbar nerve roots
entering dural sheath
and spinal nerve

Figure 55 Spinal cord and cauda equina, dorsal view of lower end

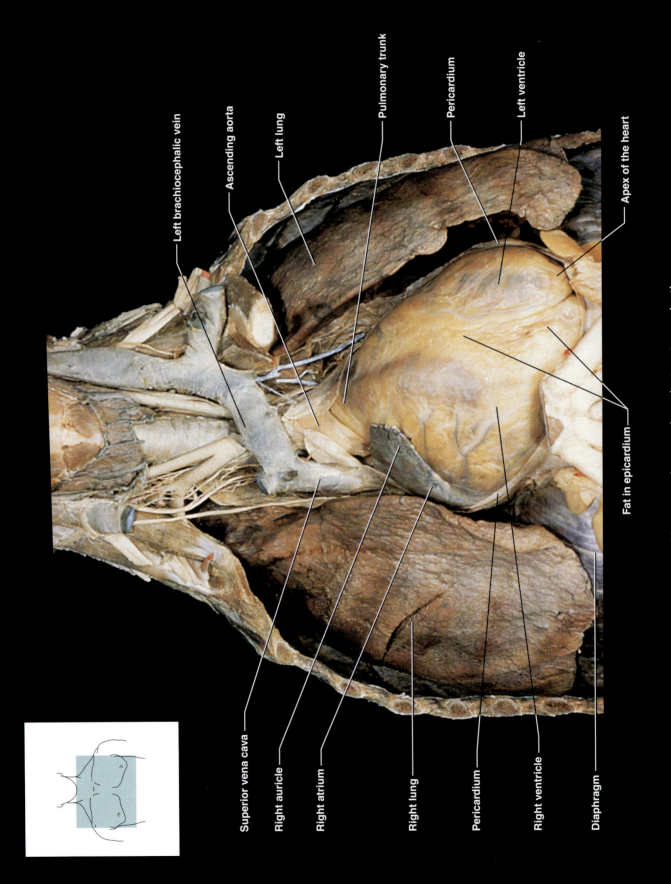

Left brachiocephalic vein

Ascending aorta

Left lung

Pulmonary trunk

Pericardium

Left ventricle

Apex of the heart

Superior vena cava

Right auricle

Right atrium

Right lung

Pericardium

Right ventricle

Diaphragm

Fat in epicardium

Figure 56 Heart and associated structures in thorax.

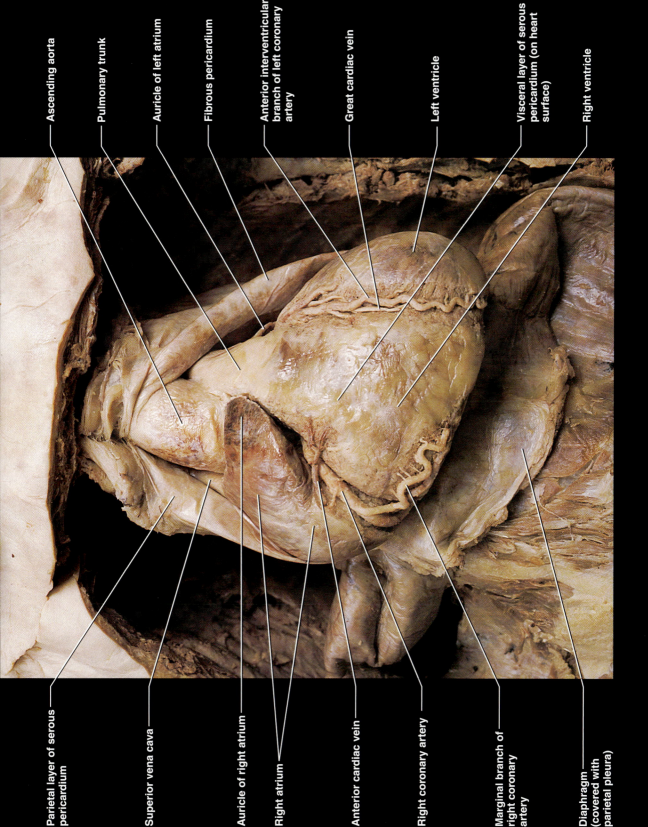

Parietal layer of serous pericardium

Superior vena cava

Auricle of right atrium

Right atrium

Anterior cardiac vein

Right coronary artery

Marginal branch of right coronary artery

Diaphragm (covered with parietal pleura)

Ascending aorta

Pulmonary trunk

Auricle of left atrium

Fibrous pericardium

Anterior interventricular branch of left coronary artery

Great cardiac vein

Left ventricle

Visceral layer of serous pericardium (on heart surface)

Right ventricle

Figure 57 Heart and pericardium, anterior view.

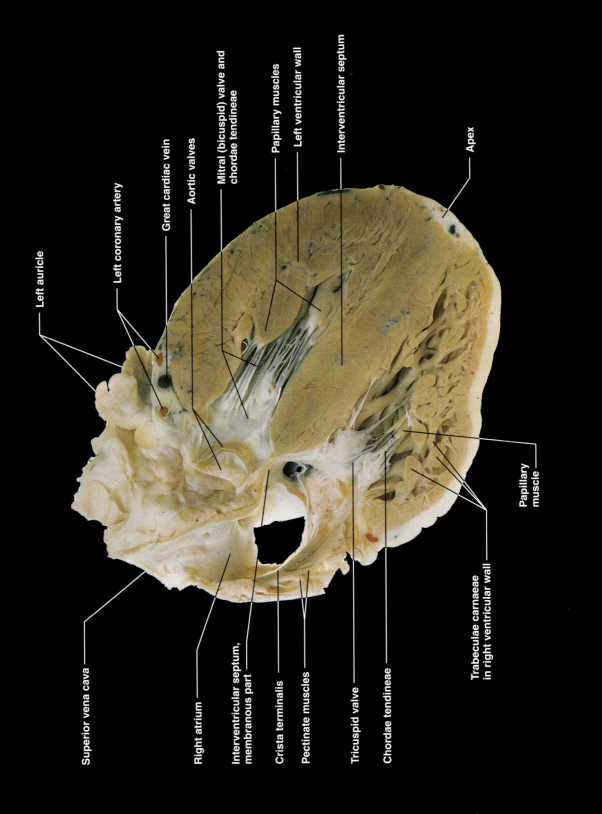

Left auricle

Left coronary artery

Great cardiac vein

Aortic valves

Mitral (bicuspid) valve and chordae tendineae

Papillary muscles

Left ventricular wall

Interventricular septum

Apex

Superior vena cava

Right atrium

Interventricular septum, membranous part

Crista terminalis

Pectinate muscles

Tricuspid valve

Chordae tendineae

Trabeculae carnaeae in right ventricular wall

Papillary muscle

Figure 58 Coronal section of the ventricles, anterior view.

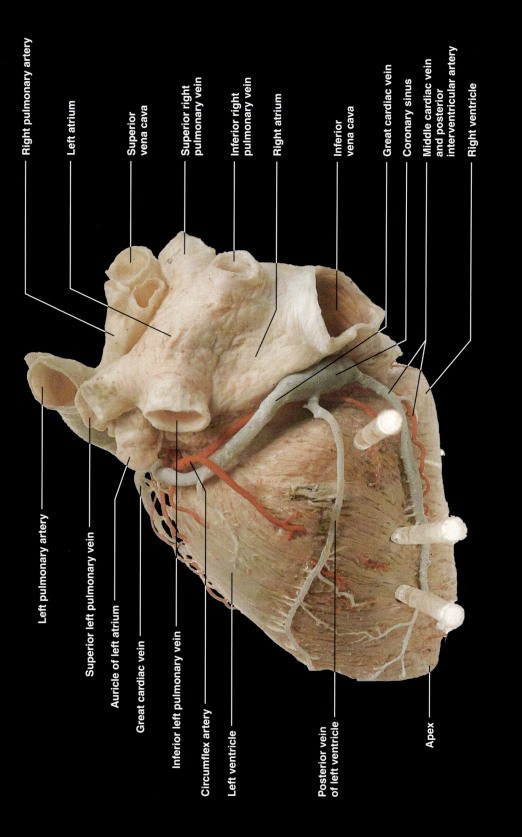

Right pulmonary artery

Left atrium

Superior vena cava

Superior right pulmonary vein

Inferior right pulmonary vein

Right atrium

Inferior vena cava

Great cardiac vein

Coronary sinus

Middle cardiac vein and posterior interventricular artery

Right ventricle

Left pulmonary artery

Superior left pulmonary vein

Auricle of left atrium

Great cardiac vein

Inferior left pulmonary vein

Circumflex artery

Left ventricle

Posterior vein of left ventricle

Apex

Figure 59 Heart, posterior view (blood vessels injected).

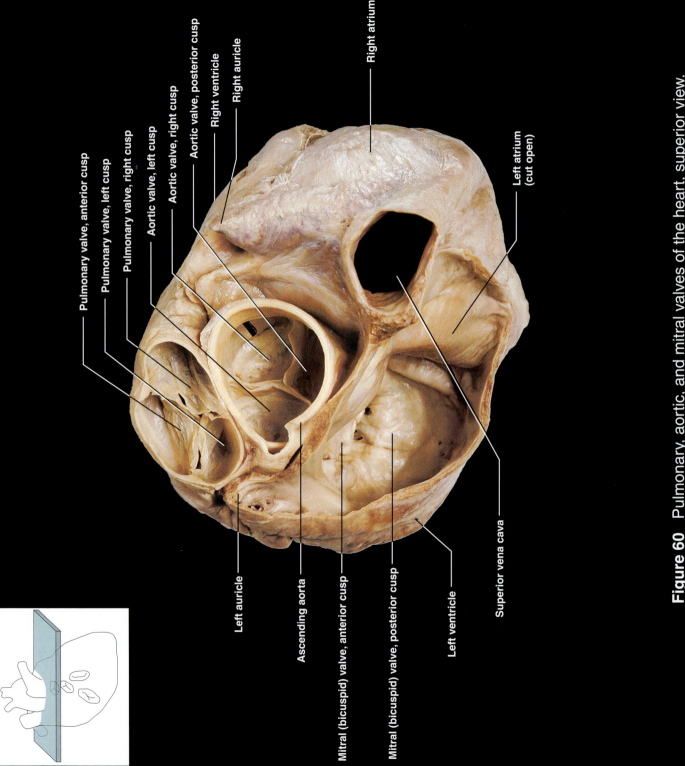

Pulmonary valve, anterior cusp

Pulmonary valve, left cusp

Pulmonary valve, right cusp

Aortic valve, left cusp

Aortic valve, right cusp

Aortic valve, posterior cusp

Right ventricle

Right auricle

Right atrium

Left atrium (cut open)

Left auricle

Ascending aorta

Mitral (bicuspid) valve, anterior cusp

Mitral (bicuspid) valve, posterior cusp

Left ventricle

Superior vena cava

Figure 60 Pulmonary, aortic, and mitral valves of the heart, superior view.

Pulmonary valve, anterior cusp

Pulmonary valve, right cusp

Aortic valve, left cusp

Aortic valve, right cusp

Aortic valve, posterior cusp

Tricuspid valve, anterior cusp

Tricuspid valve, posterior cusp

Tricuspid valve, septal cusp

Pulmonary valve, left cusp

Mitral (bicuspid) valve, anterior cusp

Mitral (bicuspid) valve, posterior cusp

Figure 61 Fibrous framework of the heart (atria removed), posterior view, from the right.

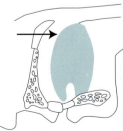

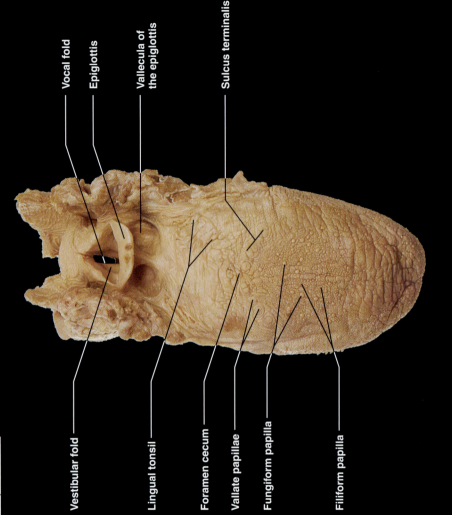

Vocal fold

Epiglottis

Vallecula of
the epiglottis

Sulcus terminalis

Vestibular fold

Lingual tonsil

Foramen cecum

Vallate papillae

Fungiform papilla

Filiform papilla

Figure 62 Tongue and laryngeal inlet.

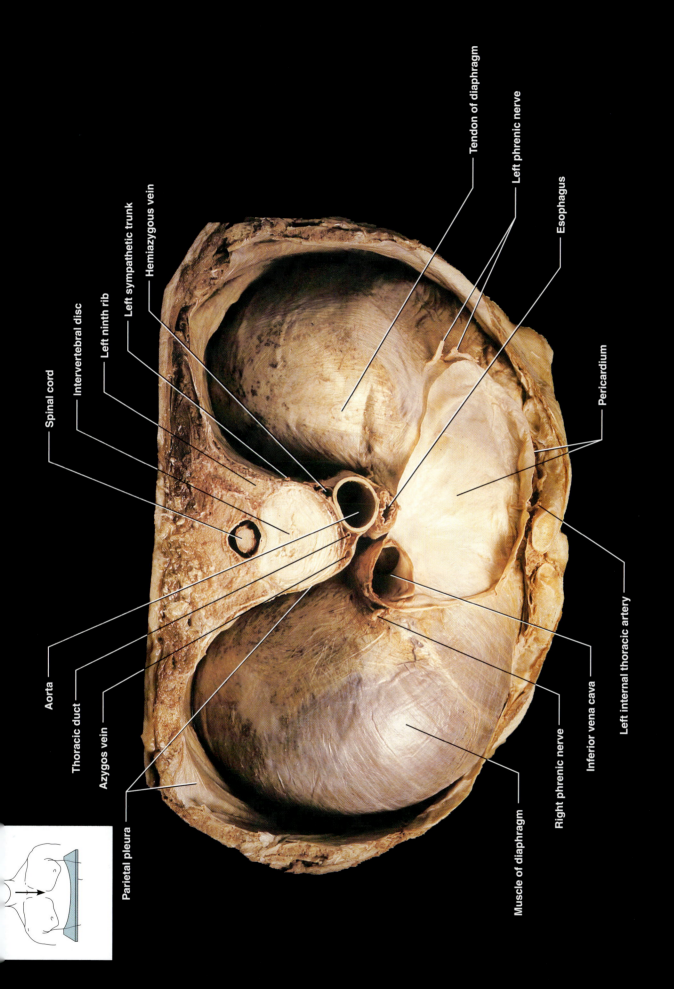

Spinal cord

Intervertebral disc

Left ninth rib

Left sympathetic trunk

Hemiazygous vein

Tendon of diaphragm

Left phrenic nerve

Esophagus

Pericardium

Aorta

Thoracic duct

Azygos vein

Parietal pleura

Muscle of diaphragm

Right phrenic nerve

Inferior vena cava

Left internal thoracic artery

Figure 63 Diaphragm, superior view.

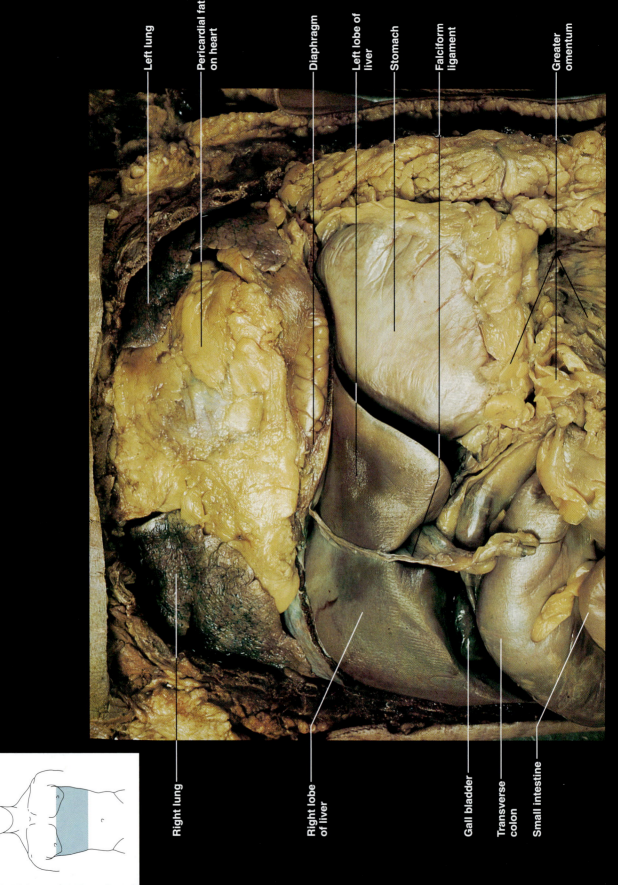

Left lung

Pericardial fat on heart

Diaphragm

Left lobe of liver

Stomach

Falciform ligament

Greater omentum

Right lung

Right lobe of liver

Gall bladder

Transverse colon

Small intestine

(a) upper abdominal viscera, anterior view

Falciform ligament

Hepatic ducts

Hepatic portal vein

Hepatic artery

Left lobe of liver

Celiac trunk

Abdominal aorta

Diaphragm

Stomach

Splenic vessels

Spleen

Tail of pancreas

Left suprarenal gland

Left kidney

Intervertebral disc

Body of vertebra T_{12}

POSTERIOR

Spinous process

Vertebral lamina

Spinal cord

Twelfth rib

Right suprarenal gland

Ninth rib

Right lobe of liver

Inferior vena cava

Diaphragm

(b) transverse section of upper abdomen, from below, showing liver, stomach, and spleen *in situ*

Figure 64 Upper abdomen.

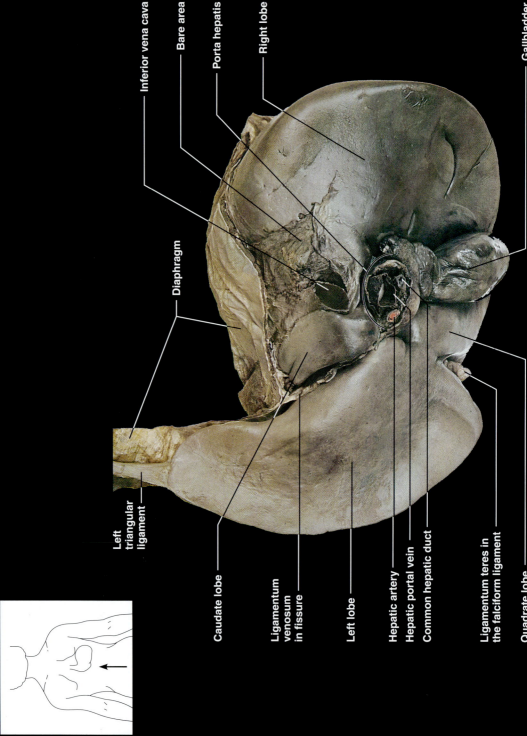

Inferior vena cava

Bare area

Porta hepatis

Right lobe

Gallbladder

Diaphragm

Left
triangular
ligament

Caudate lobe

Ligamentum
venosum
in fissure

Left lobe

Hepatic artery

Hepatic portal vein

Common hepatic duct

Ligamentum teres in
the falciform ligament

Quadrate lobe

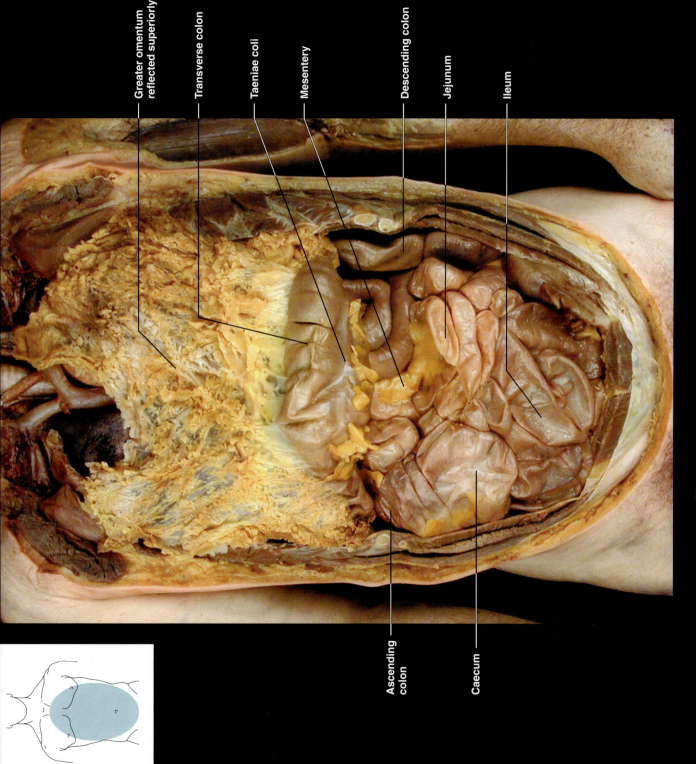

Greater omentum reflected superiorly

Transverse colon

Taeniae coli

Mesentery

Descending colon

Jejunum

Ileum

Ascending colon

Caecum

Figure 66 Lower abdominal organs.

Transverse colon

Transverse mesocolon

Descending colon

Mesentery

Sigmoid colon

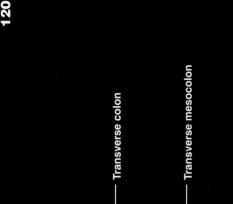

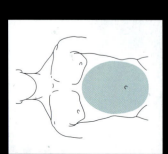

Ascending
colon

Ileocecal
junction

Ileum

Figure 67 Small intestine and colon.

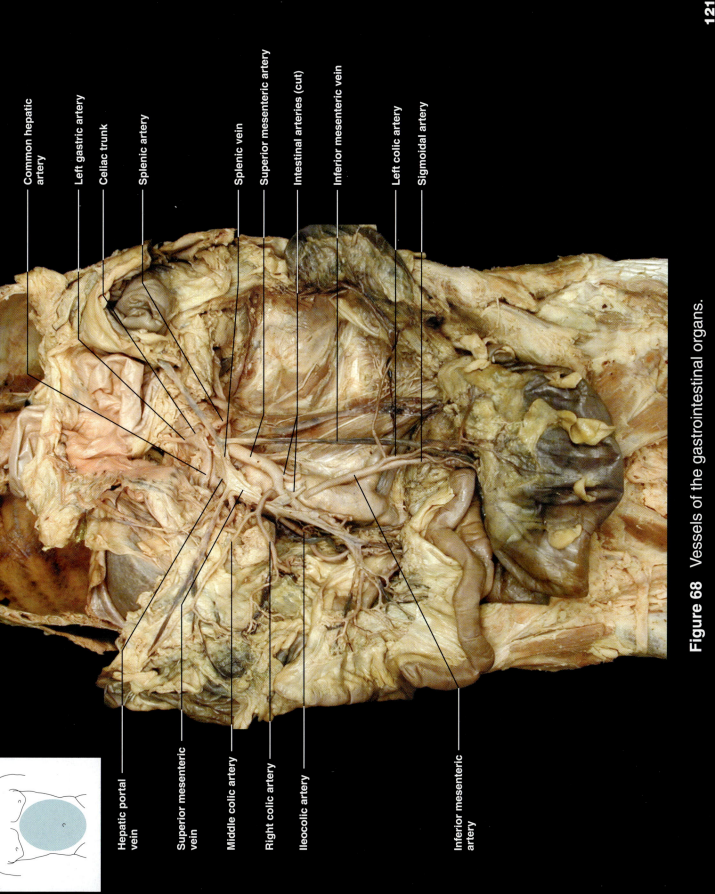

Common hepatic artery

Left gastric artery

Celiac trunk

Splenic artery

Splenic vein

Superior mesenteric artery

Intestinal arteries (cut)

Inferior mesenteric vein

Left colic artery

Sigmoidal artery

Hepatic portal vein

Superior mesenteric vein

Middle colic artery

Right colic artery

Ileocolic artery

Inferior mesenteric artery

Figure 68 Vessels of the gastrointestinal organs.

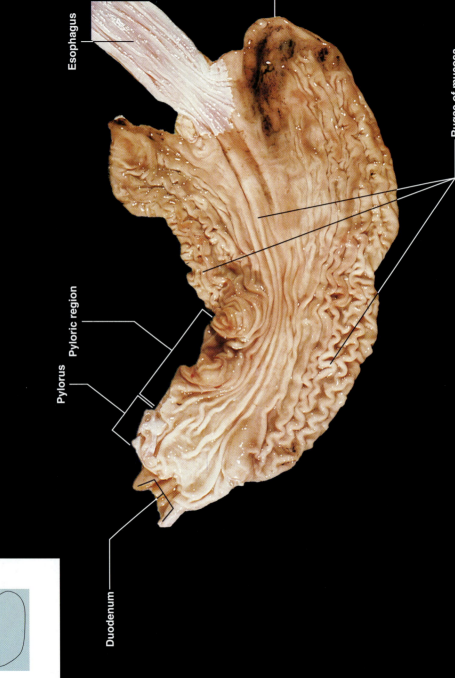

Esophagus

Rugae of mucosa

Pyloric region

Pylorus

Duodenum

(a) frontal section of the internal surface of the stomach.

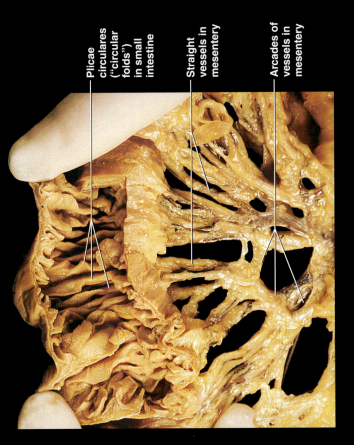

Plicae circulares ("circular folds") in small intestine

Straight vessels in mesentery

Arcades of vessels in mesentery

(b) small intestine, cut open to show plicae circulares

Figure 69 Internal surfaces of the stomach and small intestine

Abdo

Celia

Super
arter

Supra
gland

Left r
artery

Kidne

Left r

Quad

Psoas

Psoas

Ureter

Urinar

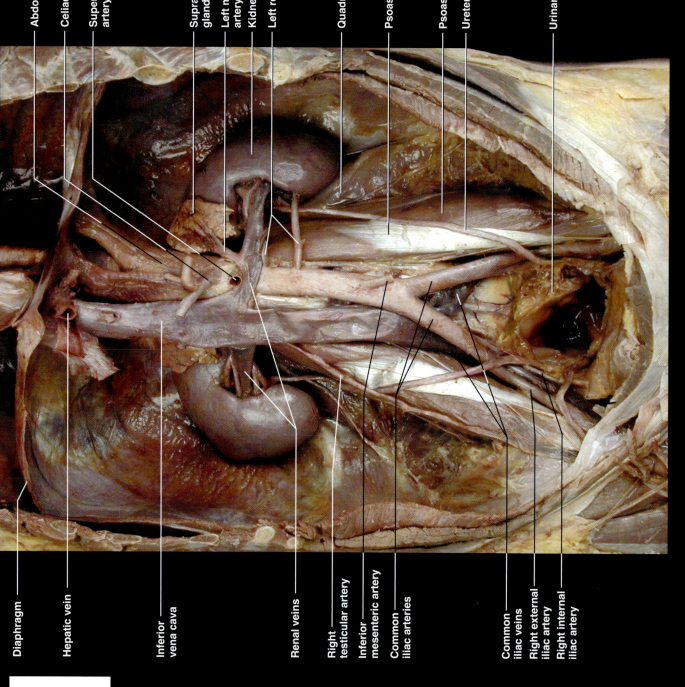

Diaphragm

Hepatic vein

Inferior
vena cava

Renal veins

Right
testicular artery

Inferior
mesenteric artery

Common
iliac arteries

Common
iliac veins

Right external
iliac artery

Right internal
iliac artery

Figure 70 Retroperitoneal abdominal structures.

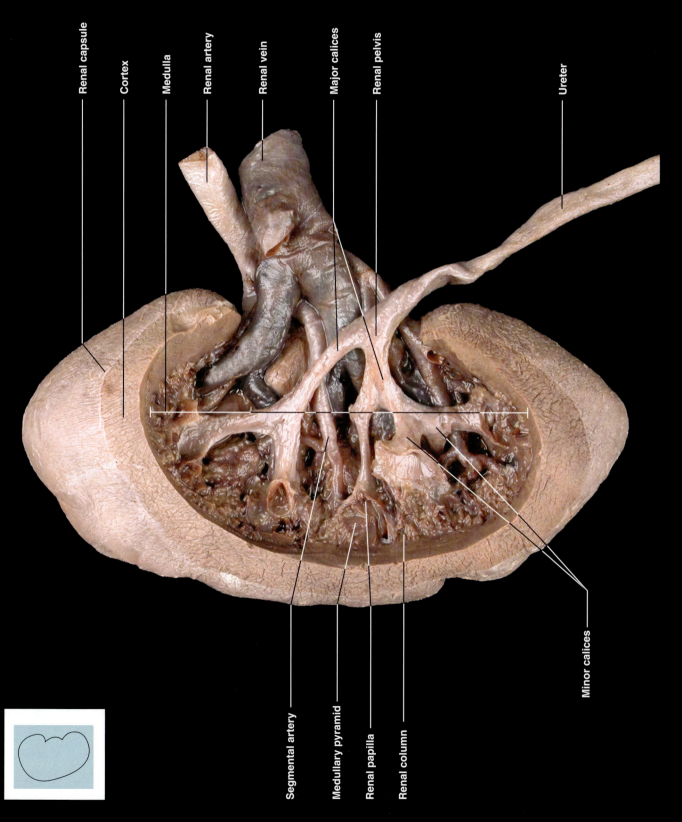

Renal capsule

Cortex

Medulla

Renal artery

Renal vein

Major calices

Renal pelvis

Ureter

Segmental artery

Medullary pyramid

Renal papilla

Renal column

Minor calices

Figure 71 Kidney, internal structure in frontal section

126

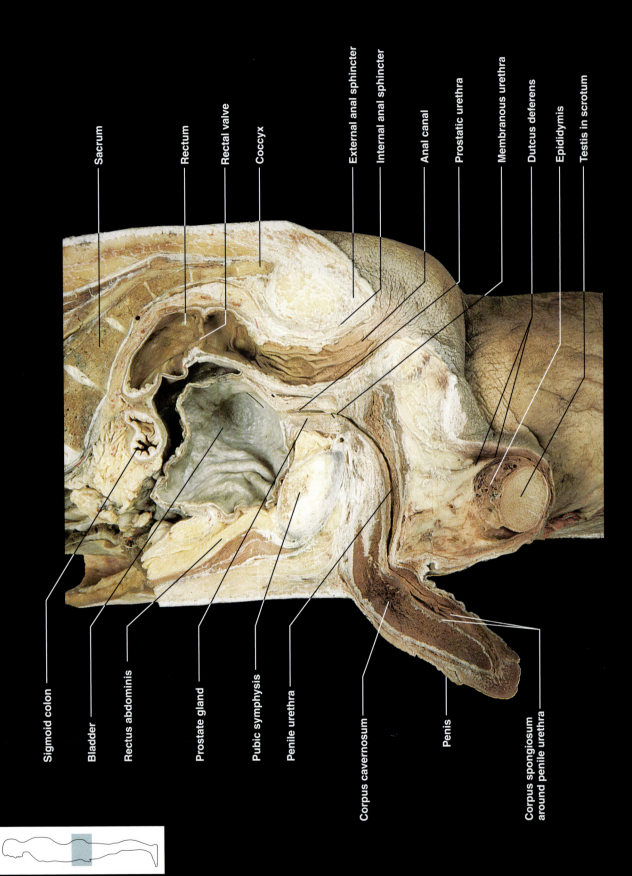

Sigmoid colon

Bladder

Rectus abdominis

Prostate gland

Pubic symphysis

Penile urethra

Corpus cavernosum

Penis

Corpus spongiosum
around penile urethra

Sacrum

Rectum

Rectal valve

Coccyx

External anal sphincter

Internal anal sphincter

Anal canal

Prostatic urethra

Membranous urethra

Dutcus deferens

Epididymis

Testis in scrotum

Figure 72 Male pelvis, sagittal section.

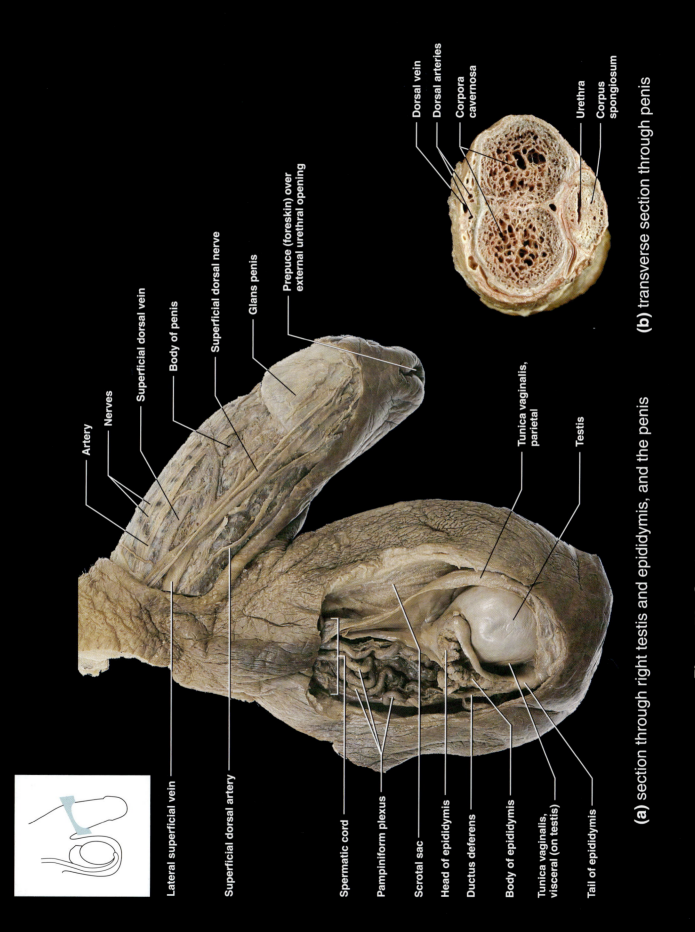

Artery

Nerves

Superficial dorsal vein

Body of penis

Superficial dorsal nerve

Glans penis

Prepuce (foreskin) over external urethral opening

Dorsal vein

Dorsal arteries

Corpora cavernosa

Urethra

Corpus spongiosum

Lateral superficial vein

Superficial dorsal artery

Spermatic cord

Pampiniform plexus

Scrotal sac

Head of epididymis

Ductus deferens

Body of epididymis

Tunica vaginalis, visceral (on testis)

Tail of epididymis

Tunica vaginalis, parietal

Testis

(a) section through right testis and epididymis, and the penis

(b) transverse section through penis

Figure 73 Sections through male reproductive structures.

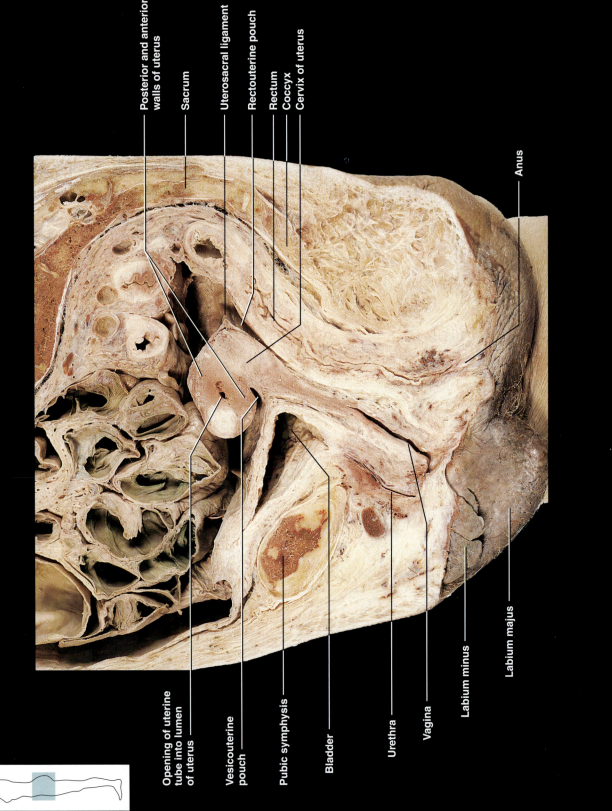

Posterior and anterior walls of uterus

Sacrum

Uterosacral ligament

Rectouterine pouch

Rectum

Coccyx

Cervix of uterus

Anus

Opening of uterine tube into lumen of uterus

Vesicouterine pouch

Pubic symphysis

Bladder

Urethra

Vagina

Labium minus

Labium majus

Figure 74 Female pelvis, sagittal section (uterus points forward in this view).

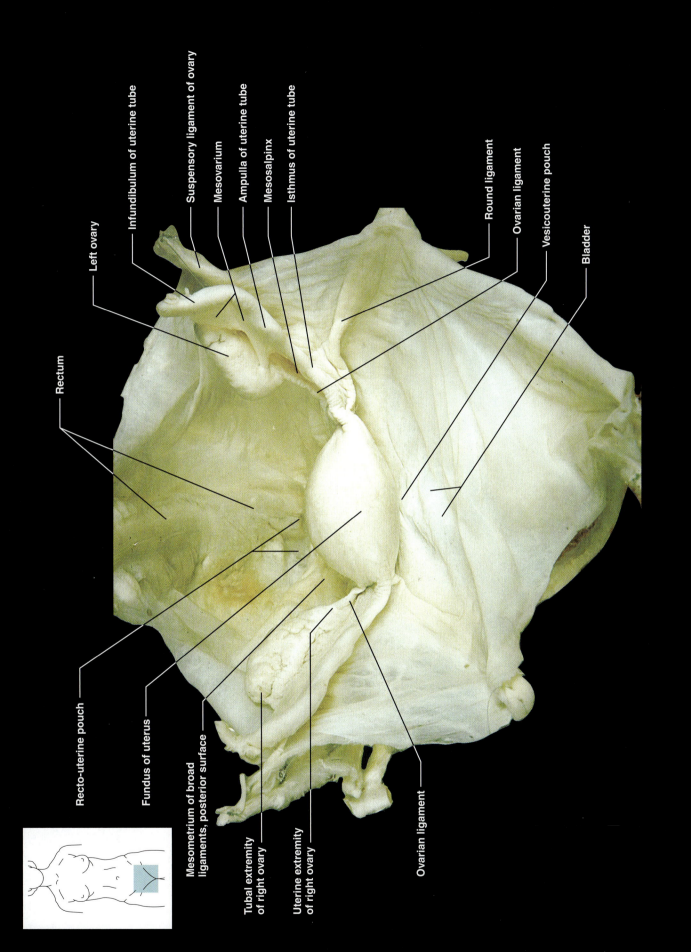

Rectum

Left ovary

Infundibulum of uterine tube

Suspensory ligament of ovary

Mesovarium

Ampulla of uterine tube

Mesosalpinx

Isthmus of uterine tube

Round ligament

Ovarian ligament

Vesicouterine pouch

Bladder

Recto-uterine pouch

Fundus of uterus

Mesometrium of broad
ligaments, posterior surface

Tubal extremity
of right ovary

Uterine extremity
of right ovary

Ovarian ligament

Figure 75   Female pelvic cavity showing the position of the uterus relative to other structures.